REBEL REFUGEES

THE CONFEDERATE EXODUS TO MEXICO

DANIEL FOXX
EDDY W. DAVISON

DRAYTONVILLE PUBLICATIONS
PEORIA, ARIZONA

Library of Congress Cataloging-in-Publication Data on file

Foxx, Daniel
Rebel Refugees: The Confederate Exodus to Mexico by Daniel Foxx and Eddy W. Davison
Includes bibliographical references and index.
ISBN 978-0-9914515-3-1

1. Confederate States of America 2. Civil War 3. Mexico 4. Reconstruction 5. French intervention in Mexico/Maximilian 6. United States of America and Confederate States of America military leaders involved in Confederate Immigration to Mexico—Biography. 7. French, Austrian and Mexican military and civilian leaders—Biography.

Edited by Mary-Helen S. Foxx
Maps illustrated by Tom Tatum.
Cover design by Ethan A. Foxx © 2015

Cover features (L to R) Maximilian I, Gen. Joseph O. Shelby, Benito Juarez and Gen. Francis Preston Blair, Jr.

This is a historical work and includes short biographies of the major participants in this story.

Published as an e-book and Printed in the United States of America
ISBN 978-0-9914515-3-1

TABLE OF CONTENTS

A NOTE ON JOHN NEWMAN EDWARDS AS A SOURCE OF INFORMATION ON THE TUXPAN, OMEALCO, AND OTHER INCIDENTS IN MEXICO

In preparing this work we consulted and studied many accounts and details of the venture to settle Confederate colonists in Mexico following the Civil War. This story has been only partially retold by modern writers largely because of the mistrust they have felt for the contemporary and memoir accounts of some who were there. It is our view that the collective story of the Confederate exiles is an important one and worth telling, therefore we must rely on these writings to present it as completely as possible, in particular the accounts of one of its most colorful participants: John Newman Edwards.

John Newman Edwards was a fighting man, soldier, writer, newspaperman, and adjutant to General Joseph Orville Shelby during the war and in the march to Mexico in 1865. As Shelby's adjutant during the war he wrote most of the general's battle reports, but he is best known for his post war literary output. *Shelby and his Men* is his memoir of Shelby's campaigns, battles and exploits during the war and a larger-than-life tribute to the general himself. He *also* wrote a rousing account of the Mexican venture, *Shelby's Expedition to Mexico, An Unwritten Leaf of the War*, that is the basis for much of the story concerning the fate of the Confederate colonies and Shelby's related exploits as a colonist.

With the failure of the Confederate colonization movement in Mexico, Edwards returned to Missouri and began collecting his memories of those momentous events that altered the lives of so many of his comrades. As Edwards was writing and publishing books based on his wartime experiences and his adventure with Shelby in Mexico, he was also contributor to and printer of the Lexington, Missouri *Expositor*. Later he was founder and editor of the *Kansas City Times*. He became well known through his writings as a defender of all things Confederate and he maintained many friendships with old comrades-in-arms and others who rode to the Southern cause. His loyalty to such men knew no limits and he became linked to some questionable characters that had worn gray. Among these were Jesse James, Cole Younger, and especially Frank James. Those friendships led to his book, *Noted Guerillas of the Civil War*, published in 1877.

Edwards is infamously accused of giving America the myth of the "noble outlaw" through this book and various newspaper articles he produced. He was known during this period for defending the James and Younger boys in their outlaw exploits, and he and Shelby even appeared as character witnesses at the trial of Frank James.

Major Edwards was a romantic of the old school. In his book, *Shelby and his Men*, he casts the wartime experiences of the Missouri Cavalry Division in terms of noble knights and gallant deeds. Written in the same vein, his record of Shelby's expedition to Mexico is as much a memoir and diary as it is pure history, but fundamentally it is considered an accurate account.

In some academic circles it is believed that his undisguised support for the Confederacy and its heroes (and villains) taints his objectivity. Therefore, many believe this perceived lack of perspective is further evidence that all his work is suspect. And though he is pilloried for his exaggeration and blatant Confederate partisanship, he was universally known as a gentleman, scholar, and a man with a high sense of honor. While it is true that Edwards wrote in the embellished prose of his age, his exaggerations were more literary than factual. But there is no evidence that he was a liar.

Besides, General Shelby might have been vain enough to stand for Edwards' flagrant lionization of his old commanding general, but Shelby would never have tolerated outright lies concerning his exploits. His vanity, in the other extreme, would have insisted that his storied exploits needed no such embellishment.

Edwards' exaggerated prose is not, however, the only reason modern historians of the Mexican venture are uneasy about citing his writings. Another principle criticism of Edwards' account is that there is a dearth of corroborative writing by other participants or observers, and most of what does exist is, like Edwards' work, difficult to substantiate. Since this is the case, Edwards' account is often the only one available.

When writing about the Confederate experience in Mexico, particularly in any detail about the collapse of the colonies and their aftermath; authors have been circumspect in relying too heavily on Edwards. The result has been that we have a skeletal coverage at best of that later period of colonization. This is not intended as a criticism of authors who have dealt with this story; they are to be commended for their work. But when Edwards is mentioned in such works their authors seem to feel impelled to give readers a cautionary note concerning the Major's reliability, as, in truth, we also seem to be doing. However, for the reasons presented here we accept Edwards' account as basically truthful, though perhaps in some parts highly colorful in its language. We liberally cite *Shelby's Expedition to Mexico* with respect to the colonies, and, when possible, support our Newman references with other citations. So, in attempting to present the most complete account of this remarkable series of events and of the men and women who lived them, we rely to some degree not only upon the pen of John Newman Edwards, but the personal integrity of both Edwards and Shelby. In so doing we believe this work presents a more complete account of the Confederate venture in Mexico, and we hope that our work does justice to both Newman and Shelby, and to all those who shared this adventure and its dangers.

INTRODUCTION

With the coming of the Civil War the young Confederacy realized that she would have to depend heavily upon foreign nations for the essentials of war and national survival. Given the overwhelming naval strength of the North, it was only a matter of time before the major Southern ports would be blockaded thus making it extremely difficult to slip through even a modicum of this vital trade from Europe.

Confederate representatives were sent to Europe in a desperate attempt to gain recognition for the Confederacy which would, it was hoped, pressure the United States government to be more prudent in dealing with the southern rebellion. Europe, however, was not their only hope for outside support. The new nation shared a common border with only one foreign country, and Confederate leaders recognized early the value they could derive from good relations with Mexico. Confederate diplomats immediately sought opportunities to not only open dialogue with their southern neighbor but to do what they could to disrupt any ties between the governments of Mexico and the United States.

Early attempts to establish friendly relations with the beleaguered government of Benito Juarez were unsuccessful, but events in Mexico were moving rapidly to a climax which would give the Confederacy new hope of Mexican aid. A short review of the history of mid-nineteenth century Mexico will shed light on the turn of events that became so significant for the Richmond government. However those in Richmond who sought to mastermind some sort of rapprochement with Mexico were poorly prepared to deal with her greatest inadequacy: national stability had always been elusive. Since the beginning of the nineteenth century the history of Mexico had been filled with disorder and violence.

In the late eighteenth and early nineteenth centuries momentous changes occurred across the globe. The Age of Enlightenment, successful republican revolutions in France and in the former British colonies in America, and the rejection of the ancient feudal system all generated a longing for change in New Spain as well. As the Napoleonic Wars wracked Europe and diverted the attention of the colonial powers from their empires, Spain in particular faced a surge of nationalism in her American colonies that led to the creation of new independent nations in South America.

The Mexican people had awakened to this nationalistic spirit sweeping the New World and launched themselves upon a long struggle for independence from Spain. The first stage of their march toward liberation began with Father Miguel Hidalgo, who issued his "Grito (Cry) de Dolores" on September 16, 1810. His call rallied peasants for the most part who were no match for the local Spanish military, and, though he recanted his actions following his capture by Loyalist troops, his revolution ended in his defrocking and execution.

Mexico marks September 16 as Mexican Independence Day, but the revolution had only begun. Another priest, Father Jose Maria Morelos attempted a short-lived revolution shortly after Father Hidalgo took up the cause of independence. The results were the same: Morelos was captured and executed, but the issue remained alive and gained more adherents.

On July 29, 1821 Mexican troops under General Antonio de Leon won a decisive battle against Spanish General Obeso at Etla. The state of Oaxaca considered itself independent from that date, though the Treaty of Cordoba granting Mexico independence from Spain would not be signed for another month. Thus ended three hundred years of Spanish colonial rule and ten years of bloody insurrection. Mexican nationalists now undertook to make Mexico a new and independent state in the world family of nations.

The first truly independent government of Mexico was not a republic but an empire that quickly followed the establishment of a military coalition in Mexico City on September 27, 1821. In an attempt to legitimize this move the new constitution created the Empire of Mexico and invited King Ferdinand VII to come to Mexico City and take the throne. If the king failed to do so he could appoint another member of the Spanish Bourbon family to the throne. If no member of the royal family was appointed, a suitable European prince could be named instead. If King Ferdinand refused to take any one of these actions then the coalition would appoint a person of their choosing to rule Mexico as emperor.

Agustin Iturbide, a criollo (creole) officer in the native Spanish army, had supported the loyalist faction until changing sides just before the formation of the coalition. He was proclaimed Constitutional Emperor of Mexico in 1822 but ruled for less than a year before being deposed. Iturbide became the first Emperor of Mexico since the reign of the Aztec kings.

The leaders of the revolution naively believed that the final act was to formally create a new national government, but they quickly learned that nation building was a difficult and often illusive aim. Spain's agreement to the Treaty of Cordoba, nevertheless cleared the way, and the first constitution of the United States of Mexico was ratified on October 19, 1824. However, when it was achieved there seemed to be no party or personality strong enough to consolidate a stable governing body. There followed one enfeebling revolution after another as leader after leader, from both the liberal and conservative camps, failed in his attempt to gain political control of the country.

Mexico was in continuous turmoil through several attempts to establish some stable form of administration and control. In the first twenty-five years after the Treaty of Cordoba two serious disruptions thwarted any attempt at stability in Mexico. The first was the war with Texas in 1836; the second was the invasion of Mexico by the United States in 1846. During that period there were at least twenty governors in the state of Oaxaca alone and nearly as many changes in the national Presidency.

Finally in 1858 Benito Juarez, a Zapotec Indian whose thorough training in the law and instinctive understanding of Mexico's political temper made him an ideal choice for the task of consolidation, was elected president of the strife-torn nation. Juarez had become a lawyer in 1834 and later a judge in 1842. He was governor of the state of Oaxaca from 1847 to 1853, at which time he went into exile partly because of his objections to the corrupt military dictatorship of Antonio Lopez de Santa Anna. Interestingly, he spent his exile in New Orleans working in a cigar factory. In 1854 he helped draft the Plan of Ayutla as the basis for a liberal revolution. Faced with growing opposition, Santa Anna, who had risen to and fallen from the presidency of Mexico several times since the 1830s, resigned in 1855 and Juarez returned to Mexico.

In 1857 the liberals promulgated a new federalist constitution and Juarez became Chief Justice and Vice-President of Mexico under moderate President Ignacio Comofort. To insure the goals of his revolution Juarez felt that a sweeping reform must be enacted.

In his view one of the fundamental problems of Mexico was the wealth and power of the Catholic Church and its vast land holdings, which the Church claimed authority to distribute to the Mexican people as it saw fit. This it did ever so sparingly, and even then the Church favored the aristocrats.

Juarez decided that this power must be removed from the hands of the clerics and invested in a government that represented the people. This was an attack not only upon the Church that reached to the Vatican itself, but an attack also upon the conservative aristocrats of Mexico. The struggle over such a fundamental issue resulted in another destructive civil war that left Juarez in control of the country, but at a terrible price. The nation was seriously weakened and there was staunch opposition from the Conservative party which attracted European attention.

By the end of 1861, the constitutional government, opposed by the Conservative aristocrats and the Church, from which Juarez had confiscated large tracts of land, found itself on the brink of still another revolution. Anarchy and chaos in the central states forced Juarez to flee to the border state of Coahuila. Pope Pius IX urged the leaders of the Clerical-Conservative coalition to oppose government confiscation of Church holdings.

The Pope was indignant at the attacks upon the Church in Mexico and perceived that the ancient institution of the Mexican Church was suffering at the hands of the Liberals. He felt driven to seek relief from whatever source he could.

Pius found the greatest show of support from Louis Napoleon III of France and seemed to give little thought to the ramifications that could result from Napoleon's opportunism. The French leader saw the distress of the Church as a reason to intervene and to use the episode as a pretext to launch a military invasion of Mexico.[1]

Exploiting old Mexican debts owed to France and other European nations, the payment of which had been suspended by Juarez as an austerity measure, Napoleon convinced Britain and Spain to join in an expedition ostensibly to occupy the Mexican port of Vera Cruz. Representatives of the three countries met in

London and signed the Tripartite Convention in October 1861.[2] The proposal, according to Louis Napoleon, was to take joint charge of the customs house at the port city until outstanding debts owed the three countries were paid.

The Spanish contingent arrived first in December 1861, joined by the French and British detachments the next month. Napoleon was anxious to bolster his ebbing popularity in France by a spectacular success in foreign affairs. The situation in Mexico seemed ideal for the French Emperor to re-establish a French Empire, which he hoped would also restore his popularity. Gradually the Spanish and British came to realize that Napoleon was trying to deceive them into cooperating in a violation of Mexican sovereignty, which would be beneficial only to France, and they withdrew their contingents the next April.[3]

The French, however, stayed and moved toward Mexico City. At the time many considered the French Army the greatest military force in the world. For nearly fifty years, since Napoleon Bonaparte's defeat at Waterloo in 1815, it had not lost a battle and had recently won major victories in both Europe and Asia. Therefore Louis Napoleon believed that if the Americans had been able to defeat the Mexican Army and capture Mexico City with only 7,000 soldiers in 1847, then certainly the French Army under General Charles de Laurencez could do the same with 6,000.

With a simplistic faith in the invincibility of the French expeditionary force, General Laurencez marched his army inland from Vera Cruz. He was not overly concerned with any armed resistance he might encounter as he approached the town of Puebla, only a hundred miles east of Mexico City.

Early on the morning of May 5, 1862 General Laurencez led his 6,000 French troops in an attack on General Ignacio Zaragoza, a Texas-born Mexican who had been ordered by Juarez to meet the French and give battle at Puebla. Zaragoza commanded 4,000 troops; many of them agricultural workers armed with only machetes and antiquated muskets. The terrain and weather favored the Mexicans in the coming fight as the recent rains had turned the uneven fields and roads surrounding Puebla into a muddy quagmire. Demonstrating his contempt for the Mexican forces, Laurencez arrogantly decided to send his troops straight into the strongest Mexican position on Guadalupe Hill. To reach the Mexican battle lines French troops would be forced to advance uphill across broken ground and ditches and through the ruins of an old adobe village.

At noon, Laurencez directed his first charge at the Mexican center, but the Mexicans held their ground and drove the French back. Although the majority of his artillery arrived late, Laurencez had the benefit of several field batteries when he made his second and third attempts to drive the stubborn Mexican troops from their position. Despite the advantage in artillery all three attacks failed, and then the Mexicans counter-attacked, aided by a force of Zacapoaxtla and Xochiapulco Indians, who though armed with ancient weapons, nonetheless succeeded in overrunning part of the French lines. Consequently, the French were humiliated and forced to fall back to Orizaba closer to the coast. President Juarez was overjoyed at the news of the victory and on May 9, 1862, he declared that the anniversary of the Battle of Puebla would become a national holiday: Cinco de Mayo.[4]

After getting their noses bloodied, the French stepped back and did what any true imperialistic super-power would do in similar circumstances: they sent for 30,000 reinforcements. In the summer of 1863 they again marched towards Mexico City, this time bypassing Puebla, for obvious reasons, and succeeded in taking the capital with little resistance.

By the end of the year the French military, joined by Conservative Mexican forces, had driven northward as far as Monterey, leaving the Juaristas in tentative control of only the extreme northern parts of the Mexican border states and those areas in the south where Porfirio Diaz, the great Oaxacan general and hero of many battles during the insurrection, held sway.

Maximilian Becomes Emperor of Mexico

With the capital and nearly three-quarters of the country under their control, the French-supported Conservatives contrived to bring a foreign ruler to Mexico, one whom they hoped the Mexicans would accept and who they planned to make a puppet prince that they could dominate and through whom they hoped to eventually restore the old order. Napoleon, who, in turn, hoped to control Mexico through such an arrangement, offered the throne of Mexico to the Archduke Ferdinand Maximilian, younger brother of Franz Josef, Hapsburg Emperor of Austria. Maximilian was a mild, idealistic man whose sensitivities were easily aroused by the pleas of Napoleon and the Conservative Mexican representatives to help bring justice and effective, beneficial government to an unhappy Mexico. Maximilian seems to have truly wanted to bring peace and prosperity to the people of Mexico, but deeply scrupulous, the Hapsburg Prince would not assume the reins of government unless the Mexican people were united in their approval of the move.

The French in Mexico, to satisfy Maximilian's desire for approval and to demonstrate the popular will, conducted a carefully controlled plebiscite, in which primarily the Assembly of Notables voted. A large majority of those questioned expressed that they wanted the Austrian for their Emperor.

Satisfied with the results of the plebiscite Maximilian accepted Napoleon's offer over the disapproving advice of his friends and family. The new emperor sailed for Mexico, arriving in Mexico City on Sunday, June 12, 1864, with his wife Charlotte, who would come to be known to her Mexican subjects and to history as the "Empress Carlota."[5]

It has been argued that the validity of Maximilian's claim to the throne of Mexico stands up to the scrutiny of history. This line of argument begins with the premise that Mexico's governmental experience from the Aztecs through Spanish colonial rule had been imperial, not republican. When Mexico achieved independence from Spain in 1821 the choice of governmental form had not been a republic but an empire marked by Augustin Iturbide's brief reign as emperor.

The French interlopers understood this in their plot to establish legitimacy for their intervention. The 1821 constitution had provided for the selection of a European prince for the throne if the king of Spain chose not to come to Mexico to rule. Napoleon's political advisors hypocritically presented what they probably thought was an unassailable candidate in the person of Archduke Maximilian von Hapsburg, a descendant of the Hapsburg Holy Roman Emperor Charles V. This ancestry was a major point in the selection of Archduke Maximilian as the ideal choice to become the Emperor of Mexico. Following this logic Maximilian was perhaps more legitimate than either the Bourbon Kings of Spain or the Mexican Emperor Agustin Iturbide.[6]

The foreign Emperor and the French troops brought more and more of the country under their control until Juarez and his supporters were reduced to little more than guerilla activity. By 1865 Maximilian's Empire was at its zenith and seemed capable, for the moment, of safely and permanently establishing itself.

Juarez, trusting in the integrity of the United States to enforce the Monroe Doctrine as soon as its own civil war was over, turned almost exclusively to the United States for moral support and occasional material aid. This, apparently, was not a sudden turn of policy on the part of Juarez. Five years earlier, in 1859, Juarez had requested a loan from the United States in return for extensive concessions against Mexican sovereignty.

Today Juarez holds a heroic position in Mexican history and his reputation is beyond reproach. However, the foundation of his government was far from being above question at the time, partly because Juarez resorted to military power to enforce the reforms he had promised as a presidential candidate. The conservative movement hotly contested the power of the Juarez government until the end of the "Reform War" when Juarez drove the Conservatives into exile.

The Conservatives and the Church did not universally accept the Juarez government. This is seen in their refusal to accept defeat by seeking support in European capitals. Even at home in Mexico overwhelming support was lacking. Before Juarez and his followers could implement the promises of the Reform his enemies were laying plans to install a European prince as emperor to replace him.

An example of this perceived discontent is seen in the popular reception given Emperor Maximilian and Empress Carlota when they arrived in Mexico City in the summer of 1864. Another argument in support of this theory is the fact that early on the French army so often had a numerical advantage on almost every occasion when they met the Juaristas in battle. At that point one would expect that Juarez should have been able to outnumber the French at every juncture, yet, for some reason a good portion of the Mexican population refused to fight for his presidency.[7]

France on the other hand, had much to gain by a Southern victory in the American Civil War and therefore supported the Confederacy. Napoleon had a significant interest in halting the growth of the United States. The North American country's rate of expansion and power had been breath-taking and threatened all other world powers. If Maximilian's reign in Mexico could lead to a successful conquest, it did not take much imagination in Washington to see the very real possibility of Napoleon marching north to aid the Confederate cause in an effort to divide the United States into two less powerful and less threatening countries.

It was through Emperor Maximilian and his French-supported government that the Confederate States of America established relations with Mexico, which proved to be so valuable to the Southern war effort. With the unofficial blessing of Maximilian and the French, a thriving trade sprang up between Texas and Mexico which brought into the South, in exchange for her most desirable product, cotton, all kinds of weapons, munitions and other supplies. Mexico had been so helpful to the South throughout the war that it should be little wonder that ranking Confederates looked to Mexico for refuge as their bid for independence crumbled around them. Maximilian's Mexico had been their friend in war; they now planned to make it their new home in defeat.

CHAPTER ONE

Confederate Relations with Mexico, 1861-1865

As early as May 1861, the Confederate Government appealed to the Constitutionalists under Benito Juarez in an effort to establish a close alliance that they hoped would open channels of supply across the Rio Grande and raise Confederate prestige in the community of nations. Juarez, however, feared Southern expansion into the northern Mexican states and the possible occupation of Monterey. He was influenced in his resistance to Confederate proposals by Thomas Corwin, United States representative in Mexico City, who wrote to Secretary of State William H. Seward in July 1861, that many well-informed Mexicans believed that independence of the Confederacy would be a signal for a war of conquest to establish slavery in each of the states of Mexico.[1]

Robert Toombs, Confederate Secretary of State, sent the tactless John T. Pickett to Mexico in the late spring of 1861, with instructions to request a treaty of friendship with the Juarez government and to circumvent any and all efforts by the United States to establish hospitable relations with Mexico.

Pickett was a Kentuckian and the ex-Chief of Staff for General Breckenridge when both served actively in the United States Army. Nearly sixty-five years old and "broken in health" he had retired from active military duty. He served as United States Consul from 1853-1858, and resigned from government service in 1861 after Lincoln was elected president in the fall of 1860. That same year Jefferson Davis appointed Pickett Commissioner to Mexico. It proved to be a bad choice. His airs of superiority and complete disregard for diplomacy, along with Corwin's effective anti-Southern propaganda, only reinforced Mexican apprehension.

As Corwin gained greater support for his government through his diplomatic maneuvering, Pickett became more indiscreet in his efforts to influence the Mexican government. Pickett's clumsy handling of his assignment antagonized Mexican government officials and weakened what little prestige his government enjoyed in the Mexican capital.

Pickett's approach to diplomacy was a combination of intimidation and bluster that cost the Confederate government any semblance of support in the Mexican seat of government. He ridiculed Juarez' authority by informing some influential Mexicans that his business was to recognize Mexico "provided I can find a government that will stand still long enough."

When Pickett began an independence movement in Vera Cruz it became readily apparent that Pickett was not going to play the role of friendly ambassador. His actions began to worry the U.S. citizens who resided in Vera Cruz and prompted United States Secretary of State William H. Seward to threaten the use of Union troops to occupy Sonora. Mexican President Juarez seemed to be agreeable to Seward's suggestion. He reacted by introducing a bill that would allow Union soldiers to move into Mexico's Northern provinces and to freely use the port of Guaymas. Though this move might have revealed some desperation on his part Juarez was willing to take whatever measures necessary to prevent the

Confederates from getting a strong foothold in Northern Mexico. The Mexican Congress passed the bill on June 20, 1861.

As for Pickett, there seemed to be no limits to which he would go to offend. On another occasion he expressed his disdain for the efficiency of the officer corps of the Mexican Army when he was told of rumors that Mexican officers might join the United States Army. He replied, "I have expressed the hope that they would all go and only regretted that the entire United States Army is not officered by Mexico."[2]

His blundering finally forced o to recall him when he voiced an absurd threat that an army of "thirty-thousand southern diplomats" might invade Mexico to prevent the Mexican government from giving aid to the United States.

Corwin used Pickett's statements for all they were worth in convincing Juarez that the Confederacy had now turned its eyes southward to satisfy its lust for territory. Pickett's dispatch relating the incident suggested that it would furnish "a golden opportunity to the people of the Confederate States of fulfilling speedily a portion of that inevitable destiny which impels them Southward."[3] Richmond had no intention of violating the integrity of Mexico and recalled Pickett immediately, but not before the Confederate ambassador spent thirty days in a Mexican jail as a result of his outrageous conduct and public statements. The damage, however, had been done. Juarez rejected the Confederate overtures and extended a friendly hand to the United States.

Corwin could not have been happier that the Juarez government was determined to deny to the Confederates the abundant and alluring riches found in northern Mexico where productive mining operations produced substantial amounts of gold and silver. The land was ripe for ranching or agriculture, and, more importantly, the numerous ports along 1,500 miles of coastline in Chihuahua and Sonora might have been used by the Confederates to beat a Union blockade of their own ports. If the Confederates could successfully occupy northern Mexico there was a real fear that it would lead to the conquest of California, possibly diverting California gold from Washington to Richmond.[4]

In spite of Pickett's blundering and the lack of support for the Confederacy in the capital, there was considerable pro-Confederate sympathy at Vera Cruz, Tampico, and in Lower California. Early in 1862, Jose Augustin Quintero, a citizen of Texas at the time, was sent by Toombs to establish good relations with Santiago Vidaurri, governor of Nuevo Leon and Coahuila, who enjoyed great autonomy by virtue of the chaotic conditions and weaknesses of the Liberal government.

Quintero was a Havana born, ex-Cuban revolutionary who later attended Harvard College and taught Spanish in Cambridge, Massachusetts. Later returning to Cuba he became a newspaper publisher in Havana supporting the Cuban revolutionary movement of 1850-1851. Although he was imprisoned and sentenced to death for his collaboration with the revolutionaries against Spain, he was able to escape and make his way to Texas where he later became an attorney at law.[5]

When the American Civil War began Quintero enlisted in the Quitman Rifles at San Antonio and traveled to Virginia as a member of the Confederate Army. In 1862 he was transferred to the diplomatic service and appointed a confidential agent, or spy, for the Confederate Government in Mexico. His assignment was to gather intelligence for Richmond concerning the potential for agriculture, mineral resources and commercial possibilities in the states of Chihuahua and Sonora. His primary contact was Governor Vidaurri, the feudal-like leader of the states of Nuevo Leon and Coahuila.[6]

So successful was Quintero's mission in 1861 that Governor Vidaurri began to provide great material assistance to the Confederacy, even offering to annex his states to the Confederacy in return for a regiment of Texas Troops and artillery. He was the only Mexican governor to ever make such an offer to the Confederacy; however neither Toombs nor anyone else ever acted upon the offer.[7]

Probably, the damage already done by Pickett's mishandling of the entire affair was irreversible. There were few options left open for the Confederacy; they could expect very little support from Juarez. Even if the Indian president had been disposed to ally his country with the Southern States, it is doubtful that he could have been of much help. Throughout this period his weak hold on the government of Mexico was no more than a "flying squadron rather than a governing body."[8]

Nevertheless, his loose hold on the Mexican states to the north was still enticing to the Confederates who sent Brigadier General Hamilton Bee, Commander of the West Sub-District of Texas, to take over negotiations in Pickett's place. Bee had spent the majority of his life in Texas, was fluent in Spanish and possessed an intimate knowledge concerning the Mexican culture. Even with these advantages, his attempts to secure a solid alliance with the Juarez regime seemed to be in vain. Soon after his arrival in Mexico he sent a dispatch to Richmond stating that Juarez had recently "decreed martial law in their frontiers, forbid the export of any article whatsoever from Texas, and closed their customs houses." The Mexican Congress had further granted the Union the privilege of transporting troops and supplies across Sonora to attack the Confederates in Arizona.

Although General Bee would go on later to command a Texas cavalry regiment during the Red River Campaign in 1864, he would for the present continue trying to establish some sort of relationship with the elusive Benito Juarez. The Mexican president, however, would grow ever closer in his posture to the Union and to American President Abraham Lincoln whom he openly admired. Lincoln returned these sentiments.

It is unclear just when Abraham Lincoln became a strong admirer and supporter of Benito Juarez, but it was obviously some time before Lincoln became a presidential candidate. The future president wrote to Juarez soon after the exiled Mexican leader was forced to flee from Mexico City in 1858, expressing "hope for the liberty of … your government and its people." The bond between

Lincoln and Juarez strengthened in 1861 as Juarez, then President of Mexico, led the "Reform War" struggles to suspend Mexico's debt to France, Britain and Spain. The growing friendliness between the two leaders becomes more understandable when seen against the backdrop of the serious political conflict both faced in trying to bring order out of chaos in their respective countries.

As political skirmishing in both countries erupted into open warfare Lincoln's political sympathies turned to open support for Juarez and opposition to the French. Lincoln and his Secretary of State, William H. Seward, saw the French intervention in Mexico as a threat to American interests in the hemisphere, but the prosecution of the American Civil War demanded all their attention. President Lincoln was left with few options and turned to providing to Juarez whatever military aid he could.

Acting on Lincoln's orders Union General Phillip Sheridan wrote in his journal that, "we continued supplying arms and munitions to the liberals (the Juarista movement), sending as many as 30,000 muskets from Baton Rouge alone." Lincoln also ordered General Grant to dispatch Sheridan to, "Concentrate in all available points in the States an army strong enough to move against the invaders of Mexico."[9] That Juarez reciprocated is clearly shown by his actions in jailing and then expelling Pickett from Mexico.

As Confederate efforts to win Juarez over became more hopeless they turned their attentions to reaching some sort of arrangement with Maximilian and the French. The French intervention and the independent posture taken by Governor Vidaurri presented the Confederate government with another opportunity to establish a favorable arrangement with Mexico. Confederate ministers approached Maximilian's government on several occasions seeking to negotiate some kind of alliance between the two nations. Both Maximilian and Napoleon realized the dangers of an open alliance with the Confederacy. Their presence in Mexico may possibly have been in defiance of the untested Monroe Doctrine, and they must have been well aware of the official attitude of the United States toward their venture. A Confederate victory would help their empire, but if they were openly allied with the South and the Union was to emerge victorious, their scheme would have little hope of success. Maximilian therefore remained aloof, though in sympathy with the Confederacy.

There were many in Lincoln's cabinet and in the United States Congress who viewed the situation in Mexico as inseparably connected to the Rebellion,[10] not only because of the efforts made by the South to obtain diplomatic recognition, first from Juarez and later from Maximilian, but also because of the trade which flourished between Texas and its southern neighbor. Although a formal alliance was never arranged with either Mexican government, a thriving trade between Texas and Mexico existed throughout the war. Carried on at a small scale at first with ambitious Mexican merchants who dissented from the policies of the Juarez government, it grew to an enormous enterprise sanctioned unofficially by Maximilian during the waning days of the Confederacy.[11]

One example of the huge profits being made by these smuggling operations can be demonstrated by a deal organized by Governor Vidaurri. Though he had been under increasing pressure from the Juaristas as the fortunes of war continued to shift both in Mexico and the United States, he continued to broker trade before being forced to flee from Texas on March 7, 1864. His last transaction was a large shipment of twelve thousand bales of cotton assuring a gross return of nearly two and a half million dollars on that one deal alone.

Almost two years before Vidaurri's final business deal the Confederacy had suffered a major calamity. Commodore David Farragut drove his fleet through the Confederate defenses of New Orleans on April 25, 1862, thus sealing the mouth of the Mississippi River and denying the South control of its last important port. Now the Mexican town of Matamoros, across the

Rio Grande from Brownsville, Texas, became the center of this trade, along with the infamous port of Bagdad at the mouth of the river.

These two wild and wooly river-port communities boasted a cast of unsavory characters to rival any infamous port in the world. Criminals of every stripe; pirates, outlaws, renegades, contraband runners, felt as much at home there as in Shanghai, Singapore or the brigand sanctuaries along the Barbary Coast. Here, in the early 1860's, life was cheap and everything was for sale. Located some thirty miles from the mouth of the Rio Grande and situated on the mud flats just across the river from Brownsville, both towns were ridden with filth and disease, especially during the hot summer months. Nevertheless, reflecting this great surge in commercial activity, the population of Matamoros rose from 6,000 at the beginning of the war to over 40,000 by 1865; and Bagdad, formerly a cluster of rough-board shacks, became a full-fledged town.[12]

Southern cotton was in demand throughout the world, and even Juarez was neutral when it came to the great benefit derived from the sale of this commodity in Matamoros.[12] Cotton was transported overland to Brownsville and to Matamoros, and valuable supplies such as food, clothing, metals and munitions came back into Texas to succor Confederate troops. Visitors to the bustling city reported that every warehouse was filled with cotton and that mountains of it were stacked on the ground and covered with tarpaulins. During the war the cotton trade brought together in Bagdad and Matamoros over 20,000 speculators from both the Union and the Confederacy, as well as from England, France and Germany.[13]

Ships from all over Europe unloaded arms, clothing and other supplies and returned to Europe with their holds full of Southern cotton. Major General Lew Wallace wrote from the mouth of the Rio Grande in March, 1865, that he could stand on his boat and count at least one hundred vessels of all kinds off Bagdad, laden with Confederate supplies. "Neither the port of New Orleans nor that of Baltimore can present today such a promise of commercial activity," he reported.[14]

General Wallace had an impressive military career having seen heavy action in both the Mexican War and the early years of the Civil War. At the close of the Civil War he sat on the court martial that tried the Lincoln conspirators and presided over the trial that sent Henry Wirz, the commandant of the Andersonville Prison Camp in Georgia, to the gallows. Wallace would eventually become governor of the New Mexico Territory and diplomat to Turkey. He may be best remembered for his famous novel, *Ben Hur*.

The illicit trade that concerned General Wallace was not limited to foreign shipping. Enterprising New England Yankees were not willing to let such lucrative business pass them by, regardless of the war. They sailed from Northern ports to Havana and left there for Matamoros with fake registry papers and flying foreign flags, usually British. At Matamoros they were loaded with Southern cotton and other goods, much of which found its way through Northern ports into Yankee mills.

These Yankee ships necessarily brought supplies required by the Confederacy: guns, ammunition, and medicine. A correspondent for *The New Orleans Times* indicated that a merchant house had received a consignment of percussion caps from New York that was sold in Brownsville and that pistols, medicine and other supplies were regularly secured from Northern ports in the same manner. A firm of merchants in New York, largely engaged in contraband trade with Matamoros, had made more money from one ship sailing in that trade than from all its other ships combined.[15]

The United States Government was understandably alarmed and in November 1863, sent Major General Nathaniel P. Banks at the head of a Union force to occupy the lower part of the Texas side of the Rio Grande. General Banks was a former "Free Soiler" who had served in the Massachusetts House of Representatives and was later governor of Massachusetts from 1858-1861.

The Confederate Trans-Mississippi Department was desperate by this time and not willing to allow this vital trade to be disrupted. With Yankee patrols blocking the lower Rio Grande the Confederates simply moved three hundred miles north to Eagle Pass. There they negotiated with the Juaristas to bring their cotton across the Rio Grande through Piedras Negras and then safely to Matamoros on the Mexican side of the river for sale to Northern and foreign speculators.[16]

This route was more dangerous because of the activities of Federal sympathizers acting as guerillas. Issuing orders to destroy any cotton which could not be safely transported to Eagle Pass, General Hamilton Bee took personal charge of the operation. He wrote to Quintero in Monterey, that he had taken over one million dollars worth of cotton from Brownsville to Eagle Pass with little difficulty from Yankee troops, "but great danger from renegades and traitors."[17] In July 1864, the frustrated Union force was withdrawn and the Confederate trade returned to the Brownsville-Matamoros crossing.

Neither were Mexican authorities by any means anxious to lose this trade. An annual revenue of $100,000 in import duties found its way into the treasury of whichever Mexican government controlled Matamoros at the time. The Confederacy, on the other hand, found that the favorable arrangement at the Mexican port gave them a valuable window to the rest of the world. Not only was it a port through which to ship her most valuable asset, but it also provided a safe channel for diplomatic correspondence and dispatches to all parts of the world.[18]

Although Matamoros was for all practical purposes an appendage to Brownsville, it was by no means a Confederate-controlled town. The United States consul "openly used his office and position to recruit troops for the United State Army" from among the Mexican population as well as Texans with Federal sympathies. The nightlife of Matamoros was enlivened by all sorts of intrigues, which had repercussions far into the interior of Mexico and Texas as renegades on both sides of the Rio Grande carrying the United States flag attacked Confederate supply wagons.[19]

A ruthless breed of bandits and desperados inhabited Texas and Northern Mexico during this period. Union agents referred to by the locals as *enganchadores*, often offered rewards of one hundred pesos in gold to Mexican, Indian, and Tejano outlaws to encourage them to attack Confederates in the border territories. Their operations required a large number of Confederate troops to remain in Texas rather than being transferred to the more vital eastern theater of operations. Their absence from the battlefields of Virginia and the Deep South was an added benefit to the Union Army.

One of the more well known of the bandit leaders who helped to tie down Confederate troops along the Texas border was highwayman Octaviano Zapata, a native of the area. Zapata conducted raids in both Texas and Mexico in search of money, arms and supplies. By the end of 1861 he was one of the most renowned renegades along the Rio Grande border.[20]

Because he received enthusiastic support from the U.S. Consul in Matamoros, Leonard Pierce Jr., and from Edmund J. Davis, a district court judge in Texas, Zapata often performed his services as if he were a semi-official Union officer. For these reasons and others Zapata's band of guerillas often flew the American Flag during their sorties against the Confederates.[21]

In December 1862, Zapata's men fell upon a Confederate supply train, sacking the goods and killing the soldiers escorting it. Less than a month later they attacked another wagon train in route to Fort Brown from Ringgold. These ambushes and others like them were instrumental in bringing about an extradition agreement negotiated in February 1863 between Hamilton Bee and Governor Albino Lopez of Tamaulipas, who was also a close associate and subordinate of Benito Juarez.[22]

The agreement was a loosely binding understanding that stated certain categories of criminals would be arrested and extradited. This included the guerilla leaders swarming the countryside and counterfeiters operating throughout the area. Some of these men were U.S. agents attempting to discredit Confederate currency, but others were opportunists who respected neither the Confederate dollar nor the U.S. greenback. Such activities were a concern to both the Union and the Confederacy.

Jose Quintero, acting as a Confederate spy and agent in Northern Mexico, also attended the Bee/Lopez meeting as an observer and later reported the proceedings to the Confederate State Department. The agreement was not an officially sanctioned treaty and the government officials in Richmond warned Quintero to avoid applying too much pressure on Mexican officials to demand extradition. They suggested to Quintero that "we have no right, in the absence of treaty stipulations, to demand extradition."[23]

Later in 1864, when the French forced Juarez into virtual exile in the Northern Provinces of Mexico, the Bee/Lopez agreement was nullified. However, Quintero brokered another informal negotiation between General Tomas Mejia, of the French and Mexican Imperial Army, and General James E. Slaughter who eventually replaced Bee as Confederate Commander of the West Sub-District in Texas. As a result Slaughter and Mejia signed another extradition settlement that did little more good than the first.

Though Confederate agents were less than successful in their diplomatic efforts the brisk border trade was a constant concern for the Union. There seemed to be no effective way short of war with Mexico to put a stop to Confederate activities in the border states. Southern cotton continued to stream out through Matamoras and a huge amount of supplies and gold flowed in. So valuable was the port of Matamoros to the Confederacy that General Lew Wallace felt compelled to write:

> Matamoros is to the rebellion west of the Mississippi what New York is to the United States…its great commercial and financial center,

> feeding and clothing the rebellion, arming and equipping, furnishing it materials of war and a specie basis of circulation in Texas that has almost entirely displaced Confederate paper...The entire Confederate Government is greatly sustained by resources from this port.[24]

To combat the renegade Zapata and others, such as Juan Cortina, the Confederacy assigned Santos Benavides, the highest-ranking Mexican-American serving in the Confederate Army, the responsibility of hunting down, capturing or killing the plague of Mexican banditos raiding along the border. Benavides had grown up in the Rio Grande frontier and had fought for the Federalists in support of the Mexican Government against the Centralist Movement. When Texas withdrew from the Union he and his three brothers enlisted in the Confederate Army.[25] Commissioned to the rank of captain in the 33rd Texas Cavalry upon his enlistment, he was immediately sent into the Rio Grande Military District where his exploits won him promotion to the rank of major.

The actual number of Mexicans and Tejanos who served in the Confederate Army can not be determined for certain, but estimates range from as few as 3,000 to as many as 10,000. Many of these Mexican Confederates were engaged in combat far from the Sonoran Desert. Approximately one thousand joined John Bell Hood's Texas Brigade and were transported by rail to the eastern front to participate in some of the most famous and vicious battles of the Civil War including Gaines' Mill, Second Bull Run, Antietam, Fredericksburg, Gettysburg, and the Wilderness campaign and at Appomattox Court House. Several other Mexican units were incorporated in General Henry Hopkins Sibley's Confederate Army of New Mexico and fought with him from Valverde to Apache Pass.

After Juan Cortina (sometimes known as the Robin Hood of the Rio Grande) attacked the county seat at Corrizo in Zapata County in May 1861, Captain Benavides drove the renegade force of some one hundred and fifty men deeper into Mexico and killed seven of Cortina's men in battle and executed eleven others who were captured.

Promoted to Major, Santos Benavides continued to command the 33rd Texas Cavalry, and with seventy-nine men again crossed the Rio Grande on September 1, 1863 and took up the trail of Octaviano Zapata's raiding party. On the next day Benavides surrounded Zapata's campsite near Mier. Santos was apparently authorized to terminate Zapata's command with extreme prejudice because after a bloody firefight, eleven of the outlaw gang members lay dead. Zapata himself was among the dead having been shot by Major Benavides, personally.[26]

As the border war expanded, Abraham Lincoln became more and more concerned, concerned to the extent that he began considering some extraordinary options. Desirous of ending the thriving trade between the Confederacy and Mexico, Lincoln eventually took steps which could enforce the untested Monroe Doctrine to rid the hemisphere of a European threat while simultaneously contemplating an amazing alliance between the Union and Confederate forces to be used in a joint invasion of Mexico to drive out the French.[27]

Throughout the war the press and a growing number in Congress who accused him of everything from incompetence to outright collusion with the Confederacy had savaged President Lincoln. As the costs in treasure and lives continued to mount he became desperate to find some honorable way to end the bloodshed of a protracted war. Even though in the face of criticism he had won reelection in November of 1864, Lincoln could see no definite end to the war as the newspapers reported the lengthening lists of dead and wounded.

His Democratic opponent in the 1864 election, former Union General George B. McClellan, had argued that the war would better be ended by negotiation. In the weeks leading up to the election Lincoln had despaired that McClellan's position would carry the vote and that a negotiated settlement of the war would leave the Union permanently divided. Preparing now for his second term in office Lincoln's administration was still laboring in the shadow of the Democrats' pro-negotiation philosophy and was, therefore, open to other solutions to ending the war, short of allowing the disintegration of the Union.

The major problem Lincoln faced in pursuing negotiations was the risk he would run in weakening his original position of preserving the Union. He could never contact Davis openly without the suggestion that he was extending diplomatic recognition to the Confederacy, which he had carefully avoided from the beginning of the war.

Nevertheless, on December 28, 1864 Lincoln met with General Francis Preston Blair, Jr. to discuss a possible secret mission to Richmond. Given the stakes of such a venture and the precarious position such a mission threatened to put Lincoln in, the subject of the meeting between Blair and Davis is obscure and sketchy at best. After discussing the enterprise with Blair the President agreed to secure safe passage for the general through the Union lines for a visit with President Jefferson Davis in the Confederate capital. Whatever the details of the conversation and the hopes for the undertaking might have been, Lincoln granted Blair permission to cross enemy lines and was apparently willing for the time being to trust in Providence.

The cover story, perhaps to be used in case news of the mission got out, related to an earlier incident known as the *Dalgrhen Raid.* In March 1864 a force of Union cavalry under the command of the erratic Brig. Gen. Hugh Judson Kilpatrick had been sent toward Richmond ostensibly to free Union prisoners of war being held at Belle Isle in the James River. Outside of the Confederate capital Kilpatrick divided his force and sent some 500 troops under Col. Ulrich Dahlgren further west of the city where they were ambushed and Dahlgren was killed.

A young boy named William Littlepage found Col. Dahlgren's body. Probably thinking he was performing a patriotic service,

Littlepage searched Dahlgren's pockets and found a pocketbook and two folded sheets of paper bearing a Union Army letterhead and containing Dahlgren's signed orders. These papers were passed up the chain of command until they reached President Jefferson Davis. The Confederate president was said to have been astonished when he read that Dahlgren's true purpose was to destroy Richmond and to assassinate Davis and his cabinet.[28]

Copies of these orders were widely published in the South and in Europe in an effort to encourage sympathy for the Confederacy. Officials in the North, as might be expected, claimed that the papers had been altered or outright forged by the disreputable Southern leaders. Whatever was the truth, the suggestion that Lincoln, or other high-ranking leaders with his blessing, had sanctioned assassination was not only embarrassing but produced wild conspiracy rumors. In the hands of the Confederate government the documents were a weapon, but if the Union government could retrieve the papers they would be able to control and minimize the rumors, if not use them for their own propaganda.[29]

Even if other purposes can be ascribed to the Blair mission, the retrieval of these papers alone would have been worth the effort. However, according to an interview General Joseph O. Shelby gave following the war (see Appendix A), and other sources, the real reason behind Blair's meeting with the Confederate president was something quite different. Whether Lincoln actually authorized the proposal Blair would present to Davis may never be known for certain. However it is difficult to imagine that Lincoln did not know the actual purpose of Blair's trip since Union Secretary of the Navy Gideon Welles provided the flagship of the Potomac fleet for the general's trip to meet with Davis. The assistance of the Federal Navy provides further evidence that Lincoln had sanctioned the mission, and that the mission transcended the retrieval of the *Dalghren Papers*.[30]

Blair was the perfect choice to represent Lincoln as a clandestine emissary to the leader of the Confederacy. He had many old Southern friends, among who were Jefferson Davis and his wife, Varina. The friendship with the Davises had been reinforced before the war began when Varina had visited Blair's home in Maryland on several occasions.

In preparation for his trip, Blair wrote two letters to Jefferson Davis on December 30, which he would present in person upon his arrival in Richmond. The first of these was for public consumption, the other for Davis's eyes, only. The public letter concerned the retrieval of the aforementioned papers; the second detailed what Blair considered to be the real reason for the conference. Following the usually friendly greetings of standard Nineteenth Century letter writing and some casual conversation, Blair laid out his plan to the Confederate President.

Although Blair made it clear that he had no official credentials from Lincoln, it is understandable why Jefferson Davis believed Lincoln was the author of the proposed scheme. The plan called for an immediate cessation of hostilities between the Confederate and Union forces and for a joint effort to invade Mexico to expel the French. Blair also told Davis that such an action could facilitate the restoration of the Union and help bring peace to a war-torn nation. Davis responded in a positive fashion:

> I have no disposition to find obstacles in forms, and am willing now, as heretofore, to enter into negotiations for the restoration of peace; and am ready to send a commission whenever I have reason to suppose it will be received, or to receive a commission, if the United States government shall choose to send one with a view to secure peace to the two countries.[31]

Blair returned to Washington on January 16 in an optimistic mood. He met with Lincoln on the 18th and showed him the letter

Davis had sent. He also gave the Union President a briefing in which he implied Davis' willingness to negotiate, but on the grounds laid out by Davis that such negotiations would be conducted between two legitimate countries: the United States of America and the Confederate States of America. However it is doubtful that Lincoln ever seriously considered Blair's proposal as a realistic plan. Some historians believe the entire affair was orchestrated by Lincoln to "undermine the Jefferson Davis administration…through a campaign of misinformation including the slight possibility that at least the remnants of their 'peculiar institution' could still be saved."[32]

Be that as it may, he could never accept the two-countries condition proposed by Davis and wrote a letter addressed to Blair, not to Davis himself because he did not recognize the Confederacy as a legitimate country. "I have constantly been, am now, and shall continue, ready to receive any agent whom he or any other influential person now resisting the national authority, may informally send to me with a view to securing peace to the people of our *one common country* (italics added)." With that declaration, Lincoln made it fairly clear that he would not engage in any negotiations with the two-countries condition still on the table.

Lincoln did, however, indicate that he supported a plan to offer appropriations of up to four million dollars (only about 15% of the actual investment made by southerners) to compensate slaveholders who would volunteer to abolish slavery. It appears that with the fall of Fort Fisher on January 15, 1865, both sides knew that slavery was doomed, and many members of the United States Congress thought Lincoln's recent considerations for proposed peace talks were unnecessary and dangerous. Neither were they comfortable with the idea of reimbursing slaveholders or any other thoughts of liberal Reconstruction theories, especially stopping short of abolishing slavery once and for all. "Many radical Republicans fretted that the meeting of the president with rebel leaders would be construed as recognition for the Confederacy." They also condemned the peace effort, especially with the war nearly won; fearing that Lincoln's actions would adversely affect congressional progress on the proposed abolition of slavery framed in the future Thirteenth Amendment to the Constitution.[33]

Whatever legalistic views Lincoln held about the fall-out from any negotiations with the Confederates, he took the most unusual step of agreeing to meet with Confederate government officials. At least in theory, a covert and unauthorized attempt by Lincoln to contact Jefferson Davis concerning a possible truce between the Northern and Southern armies offered hope. To put the best face on it he perhaps felt that if he played a tight "poker hand" he would give nothing away. If he remained alert to the possibilities he might achieve a major breakthrough that would, in the end, save lives and bring the war to an honorable end. While there is speculation in this view, it does represent the humanitarian side of Lincoln's character so lauded in American literature.

A meeting was, therefore, scheduled for February 3, 1865 at Hampton Roads, Virginia, aboard the Union steamer *River Queen*. Members of the Confederate delegation were Vice President William H. Stephens, president protem of the Confederate Senate Robert M. T. Hunter, and Assistant Secretary of War John A. Campbell, a former United States Supreme Court Justice. The "commissioners" representing the Union were Secretary of State William H. Seward and the "Great Emancipator" himself.[34]

The meeting lasted for over four hours, yet the only point agreed upon during the conference was the exchange of prisoners. On this point Lincoln agreed to have General Grant reestablish new terms for the proposed exchange. To show good faith, Lincoln agreed to the immediate release of the Confederate Vice President's nephew from a Northern prison camp in exchange for a Union prisoner held in Richmond.[35]

However futile the entire affair was, it still spawned wild speculation when the Confederate commission passed through Union lines investing Richmond. On the way to Grant's headquarters both sides came out of their trenches cheering wildly for, "Peace! Peace!"[36]

Although the Hampton Roads Peace Conference would not bring about an armistice, Francis Blair was apparently under the impression that his plan was still a possibility. Within a few days of the peace talks he contacted his cousin, General Jo Shelby, who was then leading his famed "Iron Brigade," in Missouri and Arkansas. Shelby said that Blair had offered him command of a joint Confederate/Union invasion of Mexico to drive out the French. Reinforcing Shelby's command of Confederate volunteers would be "every disbanded Federal soldier in the Trans-Mississippi department, who desired service of the kind I have indicated, [who] would have been permitted to cross over to the Confederates with his arms and ammunition. Fifty Thousand of these [ex-Union soldiers and officers] were eager to enlist in such an expedition." Blair's offer led a future Shelby biographer to refer to his subject as "Lincoln's Confederate General."(See Appendix A for entire interview with Jo Shelby).

General Lew Wallace, apparently independently, had conceived a similar plan. He received a communication from a Texas friend named S.S. Brown stating that, to a man, the Confederate soldiers were bitterly opposed to the French and that he was under the impression that by "judicious manipulation" they would, "in opposition to foreign intervention, rally under the Stars and Stripes."[37] On January 14, 1865, Wallace wrote to Grant that he believed the "rebel soldiery in Western Texas...would gladly unite with us and cross the river under the Juarez flag."[37] He therefore asked permission to be allowed to contact the Confederate commander at Brownsville to propose such a plan, expressing his confident belief that such an accomplishment would stagger the rebellion and bring the Confederacy to terms.

Much to Wallace's surprise, Confederate General J.B. Walker flatly and impolitely rejected his offer.[38] Recapping the mission in a letter to Grant, Wallace stated his suspicion that there was a secret alliance between the Texas Confederates and Mexican Imperialists "contemplating ultimate annexation of Texas and mutual support, or the support without the annexation."[39] Wallace's earlier confidence had fled him and he was now prepared to believe the worst of those men who had spurned his offer to forsake their cause.

By the time Wallace made his final report on his Texas mission, Grant had forced the surrender of the Army of Northern Virginia, and a few days later Johnston surrendered the Army of the Tennessee to Sherman in North Carolina. Since the fall of Vicksburg, little attention had been given to the Army of the Trans-Mississippi Department except by those Union forces engaged in the West. Now that Lee and Johnston had quit the field and were sending their paroled men home, notice turned west of the Mississippi in anticipation of the surrender of the fifty-thousand man Confederate army in that quarter. Through April and into May they waited.

CHAPTER TWO

Surrender or Exile

The Confederate Army of the Trans-Mississippi had not been idle since having been cut off from the rest of the Confederacy with the capture of Vicksburg. Commanded by General Edmund Kirby Smith, and staffed by scores of exceptional officers of every rank, this western army had remained in the field until the fall of their government at Richmond. When the news of Lee's surrender finally reached General Smith on April 21, there was a great deal of paralyzing uncertainty concerning the course of action he should take. Not even he, as supreme Confederate military commander west of the Mississippi was able to put direction into his army, either to surrender or merely disband.

Although the fall of Richmond came as a shock to some in the Trans-Mississippi, many others had considered that possibility beforehand and turned their eyes toward what they vaguely considered a friendly Mexico as a refuge. Southern frustration over the reluctance of the Juarez government to recognize the

Confederacy, had given way to optimism with the French intervention of 1862, and many Confederate officials enthusiastically anticipated a situation in Mexico which would lead to French recognition for Richmond. Although France never extended official diplomatic recognition, she maintained a friendly attitude toward the South.[1]

As Confederate patience grew short, some friction between Maximilian and Napoleon became apparent early in 1865. Confederate diplomat John Slidell, who was representing Southern interests in Paris, informed Richmond that he feared an "entente between the cabinets of Washington and Paris." Even if this did not come about, United States determination to overthrow the French scheme in Mexico indicated th at "the safety of the new empire is dependent solely upon our success in interposing a barrier between Northern aggression and the Mexican territory."[2]

It became clear then that the Confederacy would have to look to Maximilian himself rather than Napoleon as their best foreign friend, and that Richmond, continuing to hope for foreign recognition, considered a strong and healthy Empire essential to their own survival.[3] No matter how opposed to the Federal government the rebels might have been, however, their heritage was deeply rooted in the American tradition, and the idea of monarchy was distasteful to them. From the beginning, they sought the good will of the Franco-Mexican party with some misgivings. In August 1863, General Bee, who had communicated and visited with French officials at Matamoros and Brownsville, wrote, "the good feeling of the French and Mexicans toward us is particularly useful at this moment, as future events may render it necessary to use our powerful friends *to save us from a worse fate*." (Emphasis added.)[4]

The provincial French government, and later Maximilian, in Mexico demonstrated a friendly attitude toward the Southern States in spite of the hostile view held by Washington toward their enterprise. General Bee's special agent, A. Superviele, reported:

> General Almonte, who so far had shown so much friendship for me and good will for the Confederacy…told me to communicate to the authorities at the head of the Confederacy that if he, Almonte, had anything to do with the new government that was about to be created in Mexico, he expected to prove himself the true friend of the Confederacy, and that he would be ready at all times to do for her anything in his power which was compatible with the dignity and welfare of Mexico.[5]

General Kirby Smith had carefully cultivated this friendly attitude through his subordinate officers and special representatives to Mexico with great success. As reports from the East began, early in 1865, to bear foreboding signs of failure for the cause, he dispatched Robert Rose to Mexico City to feel out the Imperial government as to what its attitude might be toward accepting rebel refugees should the Confederacy fall. "In case of unexampled catastrophe to our arms and the final over-throw of the Government," he wrote, "it is my fixed purpose to…seek asylum in Mexico."[6]

It cannot be maintained that Smith had any knowledge of the official Confederate understanding regarding the Blair proposal, but it is a matter of record that others in the South had given thought to joining their fortunes to those of Mexico, either with the Liberals or the Imperialists, should their government fail.[7] Many had already given up the Confederacy as a lost cause, deserted their units and made their way to Mexico. At least one detachment sent to bring back a group of Mexico-bound deserters joined them instead of returning to Missouri.

Having received no confirmation that the government had capitulated following Lee's surrender, Smith felt that "the serious nature of the circumstances" justified him in communicating with the Mexican government. He instructed Rose to give Maximilian assurances that the Confederate government would be willing "to enter into a liberal agreement with . . .the Mexican Empire." It is reasonable to assume that, in the absence of official instructions from Richmond, and not knowing the status of his government, Kirby Smith considered himself the highest-ranking "legal" authority left in the Confederacy. He therefore authorized Rose to offer the services of his army to the Emperor:

> There is under my command an army of 60,000 men: of these are 9,000 Missourians, good soldiers, who have been driven from their homes, and would no doubt, upon favorable inducements as to immigration and protection being held out to them, take service with the power so favoring them. There are besides not less than 10,000 men, daring and gallant spirits from other states in this department, to whom a state of vassalage to the Federal Government would be intolerable, and who would gladly rally around any flag that promises to lead them to battle against their former foe.[8]

Smith then summoned his generals for a final meeting at Marshall, Texas, to which he also invited the governors of the Trans-Mississippi States. The conference had a two-fold objective: to ascertain the imminence of the danger and to provide against it. The conference convened in Marshall on May 13, with the governors meeting in executive session apart from the generals who met at the home of Senator Louis T. Wigfall. General Smith, for some undisclosed reason, chose not to attend either meeting but waited at the nearby home of a friend. Present at the meeting of the military were Generals Hawthorne, Buckner, Preston, Walker, Terrell and Joseph O. Shelby.

The military conference proceeded as if no one present had been informed of Smith's earlier communication to Robert Rose, in which he discussed an arrangement with Maximilian for those of his command who did not wish to surrender. Shelby, who had been watching the disintegration of the Trans-Mississippi army with disgust, had decided that his commander had lost his enthusiasm for resistance, and suggested to his fellow generals that Smith should be replaced.

Shelby was perhaps, by nature, the most adventurous of all his fellow generals and his new plan called for a coup, which would place General Simon Bolivar Buckner in command, consolidate the army on the "barrier of the Brazos River," and continue the war. Essential to his proposal was an alliance with one of the two regimes in Mexico.[9]

With such an alliance Shelby believed that the North would lose heart for continuing the struggle and pursue a negotiated peace by which Mexico and the Trans-Mississippi South, at least, could be opened up for emigrants from the Southern states east of the Mississippi.[10]

Buckner was flattered that he should be offered command of the army, but would not accept unless General Smith assented to the plan. A delegation led by General Shelby was sent to present the proposal to Smith, who concurred and wrote the order relinquishing command to Buckner.[11] Jo Shelby met with Kirby Smith alone to save the general as much embarrassment as possible. Shelby later noted that there were tears in Smith's eyes as Shelby, being intentionally forthright to cover his own embarrassment, said, "General Smith, the Army has lost confidence in you."

"I know", Smith replied.

"They do not wish to surrender."

"Nor do I." Smith agreed wearily. "What would the army have?"

"Your withdrawal as its direct commander", said Shelby, "in favor of General Buckner. Then concentration of the forces of the Trans-Mississippi upon the Brazos River and war to the hilt."

As John Newman Edwards, who was not in the meeting, described the scene, "A shadow like an eagle's wing passed over Kirby Smith's face. He arose slowly and faced the younger man who was deposing him."

"What is your advice, General Shelby"?

"Instant compliance," Shelby replied coldly.

"Very well," said Smith who moved to his writing desk and wrote the order making the change, then signed it.[12]

The conference was concluded and Smith returned to his headquarters at Shreveport. Shelby rode off under a full moon to his division camped at Stone Point near Kaufman, Texas. The other officers at the meeting took separate routes to rejoin their troops. However, the conference did not seem to make all matters perfectly clear to everyone who had been in attendance. In his memoirs General Alexander W. Terrell gives a different account of the plan decided upon by the generals:

We proposed, as a set of free companions to take the country [Mexico], and to stay there and establish our own government ...With a frankness required by the occasion, we stated to General Smith our intentions . . .with about two thousand or three thousand men we could take the country to the Sierra Madre mountains (sic), and as between the domination of the French and the Americans the Mexicans would prefer the latter. Our hope was then to offer the territory to the United States."[13]

Whatever confusion may have existed in the details, it was nevertheless a daring plan; but it was conceived too late. Time had run out.

Kirby Smith turned over his duties as commander-in-chief of the Army of the Trans-Mississippi as he had agreed to do, but kept his post as chief executive of the civilian government of the Department. He had held the latter position for four years as he kept the ports along the lower Gulf of Mexico open for blockade-runners. The bewildered and broken old soldier returned to Shreveport to find two communications waiting for him. One letter signed by the governors of every state, except for Governor Reynolds of Missouri, directed him to surrender his army immediately. The other, from Colonel John Sprague, United States Army, demanded his immediate surrender of the Army of the Trans-Mississippi to General John Pope. Since the Federals had no knowledge of the recent change in command they therefore considered Kirby Smith the proper military authority with whom to treat for surrender of the Confederate Trans-Mississippi Army.

Smith held out as long as he felt he could, complaining that the terms of surrender were too harsh. When Pope modified them,

General Smith then started for Mexico, leaving Buckner with instructions to finally surrender the Department. It must have seemed like déjà vu for poor Buckner who, on February 18, 1862, had been basically deserted by Generals Floyd and Pillow and left to surrender over 10,000 Confederate troops at Fort Donelson in western Tennessee. Again he dutifully obeyed and issued surrender orders to the various field commanders of the Department.

Waiting with his troops at Stone Point, Shelby was furious when he received orders to march his division to Shreveport and surrender their arms. Only Buckner had surrendered, he stormed, not the soldiers. Since the army would not have had time to reach the specified point of surrender, he determined to lead his division of Missourians in an attack on the Federals at Shreveport to inspire the remainder of the army to fight on. Because there was no printing shop in Stone Point and he had no other suitable paper at hand Shelby reportedly ripped a piece of wallpaper from the wall of a local store and wrote on it an impassioned order to his division:

> Soldiers, you have been betrayed. The generals whom you trusted have refused to lead you. Let us begin the battle again by a revolution. Lift up the flag that has been cast down dishonored. Unsheathe the sword that it may remain unsullied and victorious. If you desire it, I will lead: if you demand it, I will follow. We are the army and the cause. To talk of surrender is to be a traitor. Let us seize the traitors and attack the enemy. Forward, for the South and Liberty.[14]

Major John Newman Edwards, Shelby's adjutant who penned the only contemporary and encompassing history of the events to come, wrote, "Man proposes and God disposes." And so He did.

A heavy rain lasting for two days washed out the bridges on the road to Shreveport, and the division found itself mired "to the saddle girth" in mud and water. The army was beginning to surrender in Shreveport, the hour for a coup d'etat had passed, and instead of a revolutionist Shelby was about to become an exile.

CHAPTER THREE

Missouri to the Rio Grande with Shelby

Many historians consider Brigadier General Jo Shelby one of the greatest cavalry commanders of the Civil War. General Sterling Price said of Jo Shelby's last campaign in the fall of 1864, "Brigadier General Shelby, commanding [a] division added a new luster to his past fame as a brilliant and heroic officer, and without disparagement to the other officers I must be permitted to say that I consider him the best cavalry officer I ever saw."[1] Edwin C. Bearss wrote, "On a smaller scale, Jo Shelby was in Missouri and Arkansas what Nathan Bedford Forrest was in West Tennessee and Northern Mississipi."[2]

His victories at Boonville, Carthage, Cane Hill and Prairie Grove established him as one of the most dynamic and successful Confederate cavalry commanders in the Trans-Mississippi West. On October 27, 1864 Shelby won a bloody victory during the last battle of the Civil War west of the Mississippi at Second Newtonia in the Canadian River Country along the Arkansas border. It was a bittersweet victory for Shelby and it would not change, nor really slow the Union juggernaut and the smashing defeats suffered by the Missouri Cavalry Division at Westport, Independence and Marais des Cygnes in Kansas.

Sterling Price was being run out of Missouri by Sherman's forces and if not for officers such as Shelby and Major General John S. Marmaduke, leading desperate charges and maintaining stubborn rear guard actions, General Price's army might well have been annihilated. This was not the same Sterling Price who had destroyed the Union forces at Wilson's Creek sitting astride his horse as a hail of bullets singed the air around him; nor was this Price the same tenacious bulldog that held Missouri against overwhelming odds for over a year in 1861-1862. By late 1864 his health had deteriorated, partially due to obesity, and his great girth had sentenced him to traveling in an ambulance.

Things were falling completely apart after Second Newtonia and Shelby's retreat from Arkansas to Texas became a nightmarish death march. Small Pox broke out and raced among the troops like a storm driven wind. Men were dying by the score. The harried retreat through the rugged foothills of the Arkansas Ozark Mountains forced Shelby to abandon the wagon train carrying his provisions and famine began to decimate the ranks even further.

In early November 1864, the survivors of the Missouri Cavalry Division staggered into Pittsburg, in southeast Texas, just across the Arkansas-Oklahoma borders. Here they would bivouac through the rest of the winter and most of the next spring.

As the weather in 1865 began to change to spring, The Confederacy's fortunes began to change as well, and with a kaleidoscopic rapidity that challenged the senses. Lee surrendered at Appomattox Court House on April 9th, Abraham Lincoln was assassinated on April 14th, and Joe Johnston's surrender to Sherman in North Carolina followed on April 18th. On May 10th Jefferson Davis was captured with a few close associates fleeing across Georgia where he was falsely accused of complicity in Lincoln's assassination. On May 12th Kirby Smith stepped down as commander of Confederate forces in the Trans-Mississippi West, and on May 23rd and 24th, The Army of the Potomac, along with Sherman's Army of Georgia, some 200,000 strong, staged a victory march down Pennsylvania Avenue in Washington, D.C.

Shelby and his comrades had grave decisions before them. The options they faced would certainly bring hardship upon themselves and their families; only the severity of those hardships made any difference. Should they fight on or swallow their pride and surrender? Exile in Brazil, the Virgin Islands, Cuba, Europe, or any number of other places beyond the reach of the Union was a possibility. Many an unreconstructed Southerner chose one of these paths. And as for Mexico, who in the Army of the Trans-Mississippi, besides Shelby, knew of the venture Blair had proposed?

Even so, refuge in Mexico was not a new idea across the South. The Juaristas and the French-supported Emperor Maximilian, battling away for control of Mexico, offered what must have appeared as a golden opportunity to many war-weary Confederates now suffering the additional indignity of defeat. Even Lieutenant General Nathan Bedford Forrest fantasized about the conquest of Mexico with his own army, but in his brief vision he would join neither the French nor Benito Juarez.

Forrest said that with 30,000 men he could conquer the country in six months, confiscate the mines and church property, and hold possession of each state as he advanced, leaving four or five thousand men in each. He would then "open up the country to immigration, after I had given it a free government, and would get at least 20,000 people from the Southern states, besides many from Europe and the north." No one could ever say that Forrest lacked imagination when it came to planning military operations. However, Forrest's grandiose scheme was not all that far removed from Shelby's own ambitions to take advantage of the revolution occurring in Mexico.[3]

The decision to become an exile in Mexico was a monumental one for those who left their homes and the homes of their ancestors to seek refuge in a strange land. Mexico was equally as dangerous as it was exotic in 1865, but those who might choose to go there knew practically nothing of Mexico beyond unbridled fancy. The Confederados, as they would be known among their future neighbors, would be escaping from one civil war straight into the jaws of another.

There were many organized and armed groups who ventured into Mexico at the end of the war; however, none were as dramatic or filled with high adventure as Shelby's Expedition. Once it became obvious that the end had come, Shelby and his men would not march into Shreveport and surrender with the rest of the Confederates in the Trans-Mississippi West. Turning aside from his objective, Shelby marched his brigade back to Corsicana, Texas, where on June 1 he decided that he would go to Mexico rather than surrender. The Missouri Cavalry Division no longer existed as a part of an official army, but one duty remained to be done. In his final act as a brigadier general in the Confederate cavalry corps Shelby called for volunteers who felt as he did to join him in exile in Mexico. Although there were other organized and armed military groups who went to Mexico at the end of the war, an account of Shelby's expedition demonstrates the resolve shared by the Confederates who left the United States in 1865.

Major Edwards wrote that Shelby did not make a farewell speech for the history books, and indeed there is no record available of the words he spoke to his men. Thomas Westlake, one of the rebel troopers who accompanied Shelby on his expedition, kept a sporadic journal in which he wrote that Shelby expressed his fondness for his men and "requested all that would to go with him."[4] Several hundred volunteers of the old Iron Brigade joined their commander. The exact number is lost to history because the size of the expedition would vary from day to day and hour to hour as more refugees joined Shelby along the way.

The re-organization of Shelby's command was the first order of business before marching into the unknown. An election was held to choose a new chain of command. Shelby had made it clear that he was as willing to follow as to lead, but there was no other in the brigade whose leadership qualities matched his. His men unanimously elected Shelby as their colonel and commanding officer.[5] Sweeping changes were made in the brigade when the elections of subordinate officers continued. Some majors became corporals and others who held the rank of lieutenant became majors.

When the re-organization was complete Shelby turned to logistics. Equipping the expedition for the coming journey presented no problem. To Shelby's advantage, Texas was a vast arsenal of artillery, small arms and ammunition; as well as a storehouse of supplies, all of which had been run through the naval blockade to Galveston and Matamoros. Quartermaster and commissary stores existed across east central Texas and were filled with French and Austrian rifles, ammunition and cannon. Shelby chose ten new Napoleon howitzers along with six hundred rounds of shell and canister. He also commandeered 6,000 Enfield Rifles, 40,000 rounds of small arms ammunition and five hundred Dragoon Sabers. The supply was limitless as hundreds of thousands of imported muskets were warehoused in many Texas towns, stored there for the use of the Trans-Mississippi army since they could not be transported to the East. Shelby's men took what they wanted. They had never been so well armed and supplied during the war and now they were loaded for bear and ready to face what ever got in their way.

> From the colonel to the rearmost men in the rearmost file it was a forest of Sharp's carbines. Each carbine had, in addition to the forty rounds the soldiers carried, three hundred rounds more in the wagon train. Four Colt's pistols each, dragoon size, and a heavy regulation saber (sic), completed the equipment. For revolvers there were ten thousand rounds apiece. In the wagons there were powder,

> lead, bullet molds, and six thousand elegant new Enfield's (sic).

The breakup of the Confederate armies had taken away the last stabilizing influence of order in East Texas. Law and order would not return until the Federals under Philip Sheridan could assume full control. Bands of robbers, cutthroats and scavengers lost no time in terrorizing the Texas towns and countryside. Shelby swore on his honor that he would protect government property and Confederate citizens, and lectured his men on the job they must perform before crossing into Mexico.

Before breaking camp and moving southward towards Waco, Texas, Shelby called one more council of war. He hoped to stir the men into striking the enemy once more before departing. Shelby had not surrendered and he intended to fight the Yankees wherever he found them. He encouraged his men to see matters in the same light.

"Before we march southward, I thought we might try the range of our new Napoleons," Shelby said to his men. No one answered the colonel because they all knew what Shelby was thinking. The fighting spirit that drove him during the war still burned brightly in his breast. It was not in his personal constitution to turn his back on the enemy and run. He thought that if he could rally his troops and strike the Federal forces at Shreveport before the Confederates could march in and surrender, the war might be continued. There was no answer from Shelby's men; only silence from the ranks.

The prolonged silence told Shelby what he had not been able to admit, even to himself; he realized finally that the war against the Yankees was over for him. He had undoubtedly seen the reality months before but had not been able to completely let go. If he had things his way, he would have made one more attempt to prolong the war. He would have immediately thrown his brigade upon Shreveport in an attempt to seize the government and arm the Confederate prisoners being held in temporary camps. Continue the fight and never surrender. Such was Shelby's indomitable spirit but it would never contend with the Yankee army again; his duty now was to the men who stood ready to follow him into the unknown dangers beyond the Rio Grande.[7]

The brigade moved out in the first week of June. Heading south the newly formed expedition came upon an abandoned wagon train bound from Brownsville to Shreveport carrying salt pork, rice, bacon, dried fruit, whiskey and other additional provisions that delighted the troops. These supplementary rations would be more appreciated later on in the march.

As the brigade continued its march through Waco and on towards Houston they were often mistaken for another one of the numerous bands of renegades who were beginning to terrorize the countryside. Honest citizens, learning to deal with the difficulties descending upon Texas in the wake of the Confederate collapse, evacuated their path.

Texas civilians weren't the only ones trying to hold on to the vestiges of civilization and normalcy; the men of Shelby's brigade had the added dilemma of dealing with the traumatic changes the war had wrought in them. These were men who had experienced all the worst of war and its affect on them had redefined the limits of law and civilized sensibility. Though they still carried their strict sense of honor and duty they gave little thought to forgiving an insult or a slight, either real or imagined.

Before the expedition reached Houston the first of several duels involving Shelby's own men took place. Shelby was to learn that maintaining military discipline among these desperate men was going to be a complicated proposition, especially under the lawless circumstances that lay in their way. John Newman Edwards tells the story but does not give the names of the duelists, other than to say that one was a lieutenant and the other a captain. The dispute, over a woman, had come to blows and the challenge had been made. The terms of the contest were these: the two principles were to meet at daybreak, one mile from the camp. The weapons for each would be a revolver and a saber and they would begin the duel on horseback, back-to-back and twenty paces apart.

At the signal they were to wheel their mounts and fire, advancing if they so chose.

As the contest began both sat astride their horses, feet resting lightly in the stirrups, left hands holding the reins, right hands gripping their Navy Colts. At the signal, "Ready Wheel, Fire," the combatants pivoted, charged and fired simultaneously. Thundering towards each other with guns blazing the two men closed upon one another until their horses nearly collided. At powder-burn range the lieutenant fired the killing shot, which struck the young captain in the head. "A great dragoon ball had gone directly through his brain, and the captain was dead before he touched the ground. They buried him before the sun rose, before the dew was dried upon the grass that grew upon his premature and bloody grave."[8]

Even before the mission had really gotten started, blood had been spilled. Shelby's men realized that if word of the duel leaked to their commander, there would be hell to pay. Consequently, the colonel didn't hear about the incident until much later in the journey, but it was a troubling harbinger of things to come.

The road from Waco to Houston followed the Brazos River through a desolate landscape where ex-slaves wandered through unplanted fields, abandoned farms and evacuated communities. Even in this remote corner of the country, the end of the Civil War had created panic and pandemonium. With the recent breakdown of society in the region Houston was an appalling spectacle of chaos and overcrowding. In normal times the population of Houston had been approximately 10,000, but in the weeks since the war had ended refugees, renegades, and cutthroats alike had swarmed into the city. In the absence of civil order the streets were boisterous and often gave way to rioting, looting, and other illicit conduct.

Houston was also the site of vast quartermaster and commissary stores with warehouses filled to capacity with provisions, arms and ammunition, which attracted the attention of desperate citizens and criminals alike. Realizing that he and the brigade were the only reliable force in southeast Texas capable of restoring order, Shelby was quite willing to take on the task.

Word reached Shelby a few miles outside of Houston that hundreds of renegades had taken over the town. He swore that he would restore order and protect the property of the Confederate government even if it called for using extreme methods; including capital punishment. With the war over, the government could no longer control the use of the contents of Houston's warehouses so Shelby desired that the "women and children of true soldiers" should have them. Captains James Meadows and James Wood were selected to lead a light cavalry force of a hundred men to seize all the warehouses and hold their contents for re-distribution among the citizens in the most need. Shelby's orders to his captains were clear:

> Take one hundred men and march quickly to Houston. Gallop oftener than you trot. Proclaim to the Confederate women that on a certain day you will distribute to them whatever of cloth, flour, bacon, medicines, clothing, or other supplies they may need, or that are in store. Hold the town until that day, and then obey my orders to the letter.[9]

The captains could not misinterpret his next order: "Don't wait [to be attacked]. Attack first…Bullets first; speeches afterwards."

When Wood and Meadows rode into Houston they set to their task with deadly seriousness. When challenged by a crowd of nearly 2,000 stragglers and border ruffians who were looting the town, Captain Wood made Shelby's orders emphatic. On this occasion the crowd dispersed without a shot being fired and, over the next couple of days, Shelby's officers restored order and distributed the supplies to the needy. The outlaws began to disperse and ride away in search of plunder elsewhere when they came to believe that a detachment of cavalrymen was to remain in Houston to continue to enforce the law.

Similar scenes at Tyler and Waxahachie would be resolved with both imagination and ruthlessness. At Tyler they discovered seven hundred plunderers and deserters from both armies threatening the local arsenal. This time Shelby sent a detachment of approximately forty men led by Colonel Yandell Blackwell to handle the situation. Soon after their arrival in Tyler, Blackwell's men were forced to cut their way through a crowd of bandits to take possession of the armory. However, within a few minutes they found themselves surrounded by hundreds of armed men demanding the surrender of the outnumbered soldiers.

Shelby obviously held great respect for and confidence in his officers and troops. They were all veterans of the old Iron Brigade who had participated in more desperate and bloody fights than most men. The commander of the Missouri Cavalry Division did not believe that a large force was required to handle the circumstances existing in towns like Tyler or Houston. And apparently, his confidence was justified.

Accepting the challenge to either surrender the arsenal or die, Blackwell and five of his troopers seized a keg of gunpowder each and advanced fifty paces into the crowd. As they did so, each man left a conspicuous and ominous trail of black powder leading back into the building where several hundred more kegs were stored. Then each man poured out a small pile of powder on the ground in front of him and set his pistol barrel on top of the pile.

Someone in the crowd called out, "What do you mean to do?"

Blackwell answered directly, "To blow you all to Hell if you're within range while we're eating our suppers."

The five daredevils then began to prepare their evening meal as though they were dining in a garden rather an unstable munitions dump. One spark, generated from either side, would have destroyed half the town. A sudden hush fell over the crowd as they abruptly withdrew, and like that, the standoff was over and trouble was scarce in Tyler for weeks to come.[10]

At Waxahachie the situation was worse. Maurice Langhorne was sent into town with twenty-five horse soldiers. They were immediately confronted by a furious mob clamoring to get into the warehouses where the supplies were stored. This time the crowd was out for blood. Advancing towards the Confederates they suddenly opened fire on Langhorne's company. The rebels immediately returned fire and a bloody shoot-out ensued. After the first volley, the Missourians charged and closed ranks with the mob, slashing wildly with their sabers and firing their navy Colt six-shooters. In less than five minutes the grim work was done and dozens of dead bodies littered the streets of Waxahachie. Langhorne and his men held the town all that night, the following day, and the next night until the town was secured and peace was established.

Leaving more peace in their wake than they had found as they passed through, Shelby and his men continued their journey westward across the plains of the Rio Grande. When they neared the Colorado River they made camp about a mile south of Austin, the Texas capital, on June 11, 1865. By the time Shelby had reached Austin word of his vigilante actions had spread throughout the area.

The Governor of Texas, Pendleton Murrah, who was dying of consumption at the time, rode out to meet Shelby and welcome him on behalf of the great state of Texas. Murrah told Shelby that there were 300,000 dollars in gold and silver stored in the vault of the Sub-treasury in the city. Outlaw bands roaming the territory were a threat, Murrah continued, and he was concerned that without more loyal guns to defend it the Confederate Treasury and its contents were in danger of falling into their hands.

As the fortunes of war would have it Murrah's fears were all too real. On the very day Shelby arrived at Austin a band of renegade guerillas was planning a robbery attempt on the treasury.

Later that evening several of Shelby's troopers wandered into town for a taste of the Austin nightlife. Although it was early June, normally the dry season in this part of the country, the past few days had seen rain. That night, once again, the weather appeared ominous as dark thunderheads rolled in bringing more rain with them. As Shelby slept, his men in Austin were indulging themselves with tequila and women at the local cantinas.

After dark a large band of guerillas rode silently through the rain and into the state capital intending to loot the gold and silver from the treasury. With the treasury in sight their leader, the infamous Captain Rabb, was satisfied that there were no defenders in their way. Suddenly the band of outlaws charged straight down Congress Avenue thundering towards the State House where the money was stored.

Rabb was correct that no one stood ready to defend the treasury, for the moment. But several of the rebels recreating in town that night saw what was happening. They quickly mounted up and rode to notify their commanding officer.

When awakened and notified of the emergency, Shelby began to mobilize his command. They could all clearly hear the church bells of Austin ringing in the near distance. As they rode into the Austin town square, Shelby and his men could see lights in the State House blazing through the windows.[11]

Suddenly the night air exploded with the shrill blare of Shelby's bugler sounding the charge. As they approached the State House, Shelby threw out a tight cordon around the building. Inside the building the freebooters had battered down the vault door and were already within the treasury. Overconfident and obsessed with gold fever, the bandits were so focused on sacking up the American Gold Eagles and filling their pockets with silver, British Sovereigns, French hundred-franc pieces and gold ingots, that they didn't hear, or ignored, the bugle blast signaling their looming deaths.

With the deafening sound of the ringing church bells and

pandemonium breaking out all over town, it was no wonder the robbers did not hear Shelby's bugle calls. However, it did not matter. In Shelby's view these were not honorable men who deserved to be treated as soldiers, and he was not in a negotiating mood. Terms of surrender would not be discussed.

After surrounding the building, Shelby dismounted forty troopers and sent them forward. They rushed up the stone steps of the State House, while the rest of the men took up firing positions where they could clearly see their targets through the windows of the capital.

From out of the darkness Shelby's men opened fire. The marauders fell dead where they stood. So sudden and accurate was the fire from the attackers that the looters had virtually no time to return fire. A handful of the bank robbers did, however, make it to the front door where they met the forty dismounted troopers coming up the steps. Firing on the move their bullets ripped into the thieves, dropping them in their tracks. Witnesses said their blood flowed freely down the steps of the State Capital House, pooling in the street.

Reports of the action said that not a dollar was lost and not a man escaped. Inside the vault, across the wooden floor, and on the courthouse steps thousands of gold and silver coins lay scattered among the dead and dying. The shiny coins twinkling in the light of the moon as it ducked in and out of storm clouds seemed to mock the thieves. The Confederate cavalrymen dragged the bodies of the looters out onto the grass of the capital grounds.[12]

As though there had not been enough killing, "the wild night's work" was not quite finished. On their way back to camp later in the night the cavalrymen discovered another band of outlaws attempting to steal their horses. The men of the old Iron Brigade wasted no time drawing their sabers and carbines. Their sudden charge left thirty-nine raiders dead out of some fifty who had launched the horse raid. A thunderstorm broke as the slaughter was coming to an end. Lightning flashes illuminated the "bodies of the Texas horse thieves scattered in grotesque heaps throughout the camp."[13]

Following the night's battles Shelby conferred with Governor Murrah, who suggested that since the money in the treasury was Confederate property it should be divided among Shelby's men. The temptation to accept the offer must have been great. It had been months since the men had seen a paymaster. In a sense Murrah's offer was a tribute to the brigade, but there was also logic in it. What would happen to 300,000 dollars worth of gold and silver after Shelby left town? It would surely fall into the hands of other outlaws, or would be confiscated by the Union army when Sheridan's forces arrived to provide protection for the citizens of Texas.

Nevertheless, Shelby's unbending code of honor would not allow him to take even one double-eagle. "I went into the war with clean hands, and by God's blessing, I will go out of the war with clean hands," he declared. After a pause he added a little regretfully, "We are the last of the race, but let us be the best as well."[14]

Murrah was suffering greatly from tuberculosis, and probably realized that his days were numbered. He left the responsibility of office to Lieutenant Governor Fletcher Summerfield Stockdale, and then requested of Shelby that he be allowed to accompany the expedition across the Rio Grande. Stockdale would serve as governor for only two months before being replaced in August when President Andrew Johnson appointed Andrew Hamilton as provisional governor.

Shelby accepted the new volunteer and, dressed in his spotless Confederate uniform, Murrah rode out of the Texas capital at Shelby's side. Later in Monterey, the illness would kill him and the butternut uniform would become his shroud.

From Austin, Shelby's expedition marched south across the grasslands of Southeast Texas where the landscape gradually gave way to arid prairie covered with chaparral and mesquite trees. Three days ride from San Antonio Shelby's men made camp near the San Marcos River where a scout notified Shelby that bandits in the area were on a horse-stealing spree.

Anticipating an ambush sometime that evening, Shelby stationed a company of fifty men for an all night vigil near where the horses were grazing. Later that night, as predicted by Shelby's scout, a group of some eighty horse thieves raided the camp. As they came stampeding towards the herd of Confederate horses, a sudden sheet of flame from fifty Sharps carbines lit up the night like a lightning bolt. More than a few saddles were emptied with the first volley; the second volley decimated half the renegade numbers. Numerous horses and some forty men lay dead where they fell.[15]

The brigade's recent vigilante actions across South Texas had created an expectation of deliverance among the hard-pressed, honest citizens along their path. Word of Shelby's coming had reached San Antonio several days before the expedition approached the city. Standing outside the actual theaters of the war, San Antonio had escaped the ravages of the four-year struggle. Except for the activities of the lawless in town, San Antonio was unlike the other communities the brigade had rescued.

Beyond the reach of the war's hostilities San Antonio had thrived and had come to resemble a gold rush town by the end of the war. As a port of entry for freight-bearing wagons since 1861, its merchants and agents of the Confederate government had received not only munitions and other war supplies but also the finest European goods that the thriving black market could provide.

San Antonio had not, however, escaped the breakdown of law and order that followed the collapse of the Confederacy. Here, in the shadow of the Alamo, Shelby's party would find the same sort of chaos they had encountered in the smaller towns through which they had passed. But in San Antonio the mayhem and murder was on a much larger scale with determined desperados running wild through the city. With the rapid onset of land speculators, gamblers, deserters, prostitutes, and profiteers San Antonio had become a carnival of vice and violent crime. Seven men had recently been killed in the streets and Major Edwards wrote that, "Ruffians had taken possession of the city and were rioting in fashion."[16]

Trusting in the brigade's reputation for taming towns the Mayor of San Antonio, along with a contingent of desperate citizens, rode out twenty miles to meet Shelby and entreat him to come to their aid. The story was the same as before except in this case it would require a more serious effort to restore order in a town the size of San Antonio.

Shelby dispatched his most trusted officers, D.A. Williams and Alonzo Slayback, to act as Provost Marshals on each side of the San Antonio River that flowed through the city. When they arrived with approximately two hundred and fifty troopers under their command they placed half of their men on each side of the river to impose martial law. Once Shelby's men were established they immediately sent out patrols and posted guards at various street corners around the city.

Over the next twenty-four hours Slayback and Williams used a pincer action by moving their troops to the outskirts of the town, then sweeping back through the city driving the renegades, outlaws and deserters towards, and in some cases, into the river. Once the initial sweep was complete the clean up began. For the next two days order was restored at the cost of several executions by hanging and firing squads. John Newman Edwards, using his eloquently lavish prose, said of this police action, "Some men are born to be shot, some hung, and some to be drowned. Even desperados have this fatality in common with the Christians, and thus in the ranks of the plunderers there is predestination."[17]

Shelby moved the remainder of his command into San Antonio at daybreak on June 16, 1865, two weeks after Kirby Smith had surrendered the Trans-Mississippi Department. With an iron hand, Shelby had restored order to the streets of San Antonio within forty-eight hours of his arrival. The civilizing work of law enforcement had taken a heavy toll on his men, however, and they were in great need of some rest and relaxation. Many of them were entranced by the antique, half-Mexican city where the Spanish death flag had once flown above the Alamo. Shelby himself took up residence at the Menger Hotel near the Alamo Plaza where orange and lemon trees lined the streets.

While his patrols policed the city and his men recreated and drilled, Shelby took notice of several Confederate luminaries of the Trans-Mississippi Army who had arrived in town earlier. Some were passing through, some had taken up residence and others were in hiding. As it turned out, Kirby Smith was in the latter category. Fearing arrest by Union sympathizers and the wrath of the soldiers whom he had surrendered, Smith had traveled incognito all over Texas. Coming finally to San Antonio he had registered at the Menger Hotel as William Thompson. When Shelby recognized his old commanding officer, dressed in dusty civilian clothes and attempting to enter the town unnoticed, he immediately forgot his previous animosity toward Smith. In that bond of camaraderie that perhaps only soldiers truly understand Shelby was moved by the sight of the broken down old general and was filled with a sense of respect for his rank and former station. Quickly Shelby devised a welcome for Smith designed to boost his morale and restore his self-respect.

Within minutes Shelby had located Jake Connor, the brigade's bandmaster, and told him to muster the musicians and make formation in front of the hotel at sundown. As the sun began to set the band stood at attention in front of the Menger Hotel and struck up *Hail to the Chief.* A few bars were played with no sign of life from behind the curtains of Kirby Smith's window. Shelby interrupted the performance and told Lieutenant Connor to change the tune. "Play *Dixie*, Boys. If the old man were dead, that'd fetch him."[18]

Shelby was right. As soon as the lilting airs of *Dixie* began to echo through the street, Kirby Smith's curtains flew back and the window came open. General Smith was delightfully surprised at the sight of his former troops and in short order he walked through the doors of the hotel to cheers from the crowd of soldiers and citizens who had gathered. It was said that Smith had tears in his eyes when Shelby stepped up and said to him, "General Smith, you are the ranking officer of the Trans-Mississippi Department and we are here to report to you. Command and we obey; lead us and we shall follow."[19]

However, though Kirby Smith's spirits were buoyed by the demonstration of support he was nearly seventy years old and his health too poor to allow him either to join or lead any military campaign. Consequently the memorable welcome for Smith also became goodbye.

Of equal interest to Shelby was the impressive number of Confederate celebrities who were also in San Antonio and preparing to make their way to Mexico, either to find new homes or to join one of the opposing factions there. The list included such personalities as Governors Henry Watkins Allen of Louisiana, Thomas O. Moore of the same state, Thomas C. Reynolds of Missouri, and former governor Edward Clark of Texas. Representing the military were Generals John B. Clark of Missouri, Danville Ledbetter of Alabama, Cadmus M. Wilcox of Tennessee, Thomas C. Hindman of Arkansas, William Preston of Kentucky, John Magruder, and William P. Hardeman of Texas. And though Kirby Smith was also there, he was without the army he had earlier promised to Maximilian. To these were added scores of lesser lights who chose exile over surrender.[20]

Shelby allowed his men to savor the warm June days while he plotted his Mexican adventure. Over the past few days the expedition had grown to over a thousand men and rumors continued of hundreds on the way to join them. In the third week of June the brigade pulled out of San Antonio.

Another smaller party followed a few days later that included the embittered Confederate general, Jubal Early. General Alexander Watkins Terrell and Colonel George Flournoy led a third group.

The first several bands of exiles to cross into Mexico were composed almost entirely of Confederate officers and public officials. The only women who accompanied the first crossings were the wives of General Hindman, Colonel Duff and Colonel F. T. Mitchell, a Methodist preacher from Missouri who later established a 5,000-acre colony in the Rio Verde Valley of Mexico.[21]

The brigade marched on toward Eagle Pass through the outer regions of the sunbaked eastern Sonoran Desert. Distant lightning flashed silently in ragged chains against the southern horizon. One can only guess what fears weighed on their minds as they wondered if they would ever again see their families and the world they were leaving behind them.

The first evening out of San Antonio, the battle readiness of Shelby's Brigade would be tested. It was to be a reoccurring scene in coming weeks; one in which the rebels would exact a terrible toll on any who challenged them on their thousand mile trek to Mexico City.

During the day's ride several bands of scavengers had been spotted off in the distance, haunting their flanks and rear in hopes of picking up stray horses and stragglers from the column. The area was swarming with bands of night raiders and horse thieves; renegade parties who were plundering, raping and murdering their way across the countryside. However, these miscreants soon learned that the remnants of the Missouri Cavalry Division were neither helpless nor weak prey, but a dangerous scourge in their own right.

Late that night, as Shelby's camp began to settle down, his sentries and pickets spotted a band of horse thieves lurking in a large thicket of mesquite along the banks of a stream where nearby the brigade's horses were grazing. A troop of ten men led by D. A. Williams quietly dismounted, tied off their horses, and began to stalk the interlopers. Armed only with their Navy Colts they flanked the enemy, and on Captain Williams' signal, charged forward with pistols blazing. The combat was savage, conducted at close quarters with revolvers and Bowie knives. The bloody work was finished within minutes. Twenty-seven of the bandits lay dead alongside three rebel troopers.

Some time around June 18, two of Shelby's scouts reported to him that about three miles to their rear a Federal force of over three thousand troops were shadowing the Confederates. Old habits die hard and Shelby would relish one more opportunity to strike at the enemy he had fought for over four years. The Federal column was led by Colonel A. Johnson, and may have been trailing Shelby on orders from his commanders to stop the Confederates from crossing into Mexico.

Assuming that this might have been true, Shelby wrote a short message and handed it to Major James Mouland, his former adjutant. "Take two soldiers with you," he said, "ride to the rear…seek an audience with Colonel Johnson, and give him this."

If Johnson was looking for a fight, Shelby was more than willing to oblige. There was an obvious challenge in his communiqué:

> My scouts inform me that you have about three thousand men, and that you are looking for me. I

> have only one thousand men, and yet I should like to make your acquaintance. I will probably march from my present camp about ten miles further today, halting on the high road between San Antonio and Eagle Pass. Should you decide to pay me a visit, you will find me home until the day after tomorrow.[22]

No record exists of Johnson's reaction to Shelby's feisty challenge, but he seemed to have practiced the better part of valor and withdrew. Nor is there a record of how Shelby's officers and men felt about the possibility of delaying their march into Mexico to fight a battle with Yankee troops several weeks after all other Confederate armies had surrendered. As it happened, the whole affair came to nothing. The incident was the last between Shelby and Northern forces; that part of the war was now behind them.

The desolate beauty of the Sonoran Desert and its great emptiness of long horizons must have seemed a strange and harsh environment for men who had lived most of their lives in the rich, green Ozark region of Missouri and Arkansas. The familiar colors of their homeland were seen now only in those few willows and delicate plants that grew along the banks of the streams they had to cross. They forded the Rio Medina and passed through Sabinal and Uvalde where the desert was in bloom with alien vegetation. The treacherous spines of the Spanish Dagger plant were abundant as were the greasewood, cat claw, and cottonwood trees along the riverbanks.

For the next several days in late June the weather grew to be even more inhospitable and hot. Dehydration became a serious problem as the column of sunburned phantoms rode deeper into the wasteland with their eyes now set on the border town of Eagle Pass and the Rio Grande beyond.

Still a few days' ride from Eagle Pass the expedition rode

through a village that had recently been attacked and plundered by a band of outlaws. The pitiful villagers pleaded with Shelby to help return their stolen animals, property, and kidnapped women. Another less patient or merciful commander might not have cared enough for the indigenous people in the area to have weakened the strength of his overall command by risking the lives of his men.

Shelby apparently had meant it, and he may well have been right, when he stated in Austin, "We are the last of our kind, let us be the best as well." Shelby ordered Colonel Alonzo Slayback to pick two hundred volunteers and track down the marauders. As always his directions to Slayback were concise and to the point: take whatever action necessary to rescue the kidnapped victims and recover the stolen property. The captain had been with Shelby long enough to know how to read between the lines of such an order. He would shoot first and ask questions later.

When Slayback found the plunderers' camp his practiced eye evaluated the situation in an instant. The looted goods were stacked chest high in the middle of the camp; the defenders numbered about two hundred men. He ordered the charge immediately. Without hesitation the cavalrymen obeyed. Over and through the enemy they rode, their Colt six-shooters at work. The unprepared bandits gave way before the vicious onslaught as the survivors scattered into the hills leaving the bodies of half their comrades strewn around the camp. Just how one-sided the skirmish had been was demonstrated when Slayback's company returned with the rescued captives and stolen property without having lost a single man.[23]

The brigade was within two days march of Eagle Pass when Kirby Smith and others who had joined them broke off from the main force hoping to make better time with a smaller, faster moving group. Shelby and Smith would never see each other again. The brigade crossed the Nueces River and Turkey Creek. Before them was the Rio Grande, the last natural barrier between the country of their births and Mexico which could, at best, offer an uncertain life of exile

A day or two later they reached Paso del Auguila, the Pass of the Eagle, and rode by abandoned Fort Duncan, its tattered rebel flag still fluttering in the wind as if by habit. The Confederates had built the fort in 1861 never guessing that it would serve for so short a time.

Eagle Pass was nothing more than a few adobe huts and ramshackle buildings sitting atop a bluff on the Texas side of the river. The Mexican village of Piedras Negras, *the place of black stones*, sat just across the Rio Grande where some two thousand Mexican troops waited impatiently to learn whether the rebels were friends or foes

The ford at Eagle Pass had been a familiar river crossing for North Americans for many decades. Volunteers had entered Mexico at this point to support the Mexican war for independence from Spain, and then came American soldiers to fight Mexico in 1846. Filibusterers had also used this crossing in their efforts to create small empires for themselves out of Mexican territory.

The night of July 3, 1865, the rebel brigade camped for the last time beneath their battle flag on American soil. Shortly after daybreak on the fourth of July, Shelby and his men broke camp and moved down to the river's edge. There they paused long enough to conduct a brief, reverent ceremony. The battle flag of the Missouri Cavalry Division was lowered for the last time, weighted with stones and sunk into the waters of the Rio Grande by Captains Gordon, Elliot and Slayback. Within its folds Shelby placed the black plume from his campaign hat.[24] The brigade then crossed into Mexico with the vague hope that they would somehow find sustenance for their lost cause in a new country.

The brigade's former countrymen were celebrating America's eighty-ninth year of independence from Great Britain as they emerged from the waters of the Rio Grande. To cover his crossing

Shelby lined up his howitzers, dismounted a portion of his troops to form skirmish lines to his front and ordered cavalry detachments to both flanks. These battle preparations were all accomplished in plain view of the Mexican military garrison at Piedras Negras as a demonstration of the brigade's strength and bargaining position.

However, neither Shelby nor Andres Biesca, governor of the state of Coahuila and thus commanding the Juarista forces in Piedras Negras, made a hostile move. Shelby calmly left his troops in position and began to set up camp as he weighed his next decision: attack or negotiate. He believed that the men of the expedition, having given up their own country to serve as a company of free companions, should share in whatever choices lay before them. Though he would keep his personal choice to himself for the time being, Shelby saw clearly there were only two options: join the Juaristas and drive out the French as originally proposed by Frank Blair, a cousin of Shelby's, or to offer their services to Emperor Maximilian.

He would later say (see Appendix A) that he was attracted to the cause of Benito Juarez and the Liberals for a powerful personal reason; he saw their cause as more compatible with the American dream of freedom and self-determination. Shelby's instincts told him that Blair's original suggestion offered the best chance for success, and the proposal was open-ended enough for interpretation. As events would suggest during Shelby's march toward the Rio Grande, Blair's proposal was also common knowledge to many Union officers in Shelby's wake. By the time he reached Eagle Pass, Shelby had received over two hundred messages from Federal military groups, large and small, entreating him to wait for them at the Rio Grande. Apparently they planed to join him in a new war in Mexico against French Imperialism:

> On my march south from San Antonio to Piedras Negras I received no less than two hundred messages and communications from representative

> Federal officers, begging me to wait for them beyond the Rio Grande...Fifty thousand of these were eager to enlist in such an expedition.[25]

Governor Biesca watched Shelby's arrival with mixed emotions. He knew that the U.S. Government supported Benito Juarez, but he could not be certain of Confederate intentions. The Governor received Shelby kindly, holding out hopes that these well-armed troops might join Juarez against the French.

A meeting was arranged between Shelby and Biesca, which lasted for two days and during which the two had what diplomats like to call a *frank exchange of views*. Now that the war between the Union and the Confederacy was officially over, Shelby pointed out that the United States Government would not tolerate the French violation of the Monroe Doctrine. He also informed Biesca that General Grant had already ordered Phil Sheridan, to be ready to move into Mexico with fifty thousand troops if it became necessary to force the French out. Shelby said that Biesca should consider recruiting former Confederate soldiers into the Mexican Army to forestall an invasion by the Union Army.

> In this they will have the blessing of General Grant, who commands all things in the United States now and would rather win an undeclared war [fought by Mexicans and Confederate allies to expel the French] than a declared one. I have 1,000 men, and the brigade could soon be recruited up to division strength. And not just of disbanded Confederates, but discharged Federals, will join you.[26]

Upon hearing Shelby's proposal Biesca must have felt a sense of optimism. The Juarez government was shaky at best. Rumors arrived almost daily of another defeat at the hands of the French, Benito Juarez was hiding in the United States, and the steadfast

Porfirio Diaz was cut off in the South. What Shelby proposed was a rarely seen ray of hope for the Jauristas.

Finally the Mexican governor made Shelby an incredible offer: a ranking position in the service of President Benito Juarez and military command in the states of Nuevo Leon, Coahuila and Tamaulipas. He also suggested that Shelby set up headquarters at Piedras Negras where he could raise an even larger army of volunteers; then march on and attack the French forces at Monterey. Shelby maintained a poker face as he explained to Biesca that a vote among the men must be taken and that the majority decision would determine action for all.

Nevertheless such an offer must have been almost irresistible to an adventurer such as Shelby, who perhaps harbored personal dreams of a new Confederate Empire in Northern Mexico. Even ideas of conquest might have crossed his mind should Biesca's promises fall short. These concerns and others weighed heavily on his mind as he returned to camp to lay the Mexican offer before his men. Later that night Shelby laid out their options and his endorsement:

> If you are all of my mind, boys, and will take your chances along with me, it is Juarez and the Republic from this on until we die here, one by one, or win a kingdom. We have the nucleus of a fine army—we have cannon, muskets, ammunition, good prospects for recruits, a way open to Sonora, and according to the faith that is in us will be the measure of our loss or victory.[27]

In conclusion, Shelby added, "What he [Biesca] fails to perform for us we will perform for ourselves." He had always believed that with twenty thousand men he could take Northern Mexico and hold it. But for some reason lost to history, Shelby never fully explained to his men the full details of Blair's proposed joint invasion, or the hundreds of memos and messages he had received from Federal troops hoping to join him. Perhaps Shelby could foresee that Maximilian's empire would fall under the pressure, which could now be applied by the United States.

A council was held the next morning at which all the Confederates were invited to speak. Upon hearing Shelby's proposal, the men and officers conferred together and took a vote in which they decided that service under the Empire was preferable to throwing in with Juarez' vagabond government. After all, Juarez had extended support to the United States, while the French and Maximilian had showed preference for the Southern cause. The decision might have made sense to them under the circumstances, but seen in retrospect, it was probably the wrong one.

Alonzo Slayback and Ben Elliot were chosen to present the decision of the men to Shelby. "General, if you order it, we will follow you into the Pacific Ocean," said Slayback; "but we are imperialists, and would prefer service under Maximilian."

For some reason, they failed to see what Shelby saw. Should the United States chose to invade Mexico to drive out the French, these beloved men of his would once again be at war against their fellow countrymen.

When Shelby received the results of the vote he must have been stunned. However, he reflected for only a moment before answering. These men had stood by him and had followed his every order during the war. And even now as exiles they were still willing to follow his orders if he chose to veto their vote. He couldn't bring himself to do it.

> If that is your answer, it is mine too. Henceforth we will fight under Maximilian. Tomorrow at 4:00 P.M. in the afternoon we commence to march for Monterey. Let no man repine. You have chosen the Empire, and perhaps it is as well, but good or bad, your fate shall be my fate and your fortune my fortune.[28]

CHAPTER FOUR

From the Rio Grande to Mexico City

Governor Biesca was visibly disappointed when informed of the Confederate soldiers' decision, but he made no argument. There was nothing Biesca could have done to keep Shelby from marching off to join the French, so he made the best of the situation. Shelby knew something of the terrain ahead and that the heavy guns and supply wagons would become a burden to the brigade as soon as they left Piedras Negras. The desert they were about to cross was as formidable as any in the world and it was marked with arroyos, canyons and heartbreaking mountain ridges. The wagons were abandoned before the march began and necessary supplies were loaded in saddlebags and aboard pack animals.

Biesca offered to buy Shelby's cannon for $16,000 in silver and $40,000 in what became worthless Juarez scrip. Both leaders realized that the Confederates would need money on their journey and this transaction would solve at least one of Shelby's problems.

One might question Shelby's decision to sell his artillery to the Mexicans, who might later use his own guns against him. George Creel, then a cub reporter for the Kansas City *World,* asked Shelby that exact question in 1896. In the newspaper article that followed, Creel wrote, "But, General …the Juaristas outnumbered you five to one, according to your own admission. And you just told them you were going to join the enemy. When they bought your ammunition, guns, and artillery, what was to prevent them from taking back the money and annihilating you?"

Shelby's answer displayed the confidence and bravado of a man who knew he led one of the most mobile and fierce fighting machines ever to have crossed the Rio Grande. "You seem to forget, Sir, . . .we still had our side arms."[1]

Not only did Shelby and his men have their side arms, their fighting spirit was, if anything, even more heightened. The war had worn down their civilized sensibilities and their anger was always a short fuse away from flaring up at the slightest provocation or insult. This violent potential would be loosed many times on the march to the Valley of Mexico, but for now it flashed to dangerous life in an incident involving horses.

Along the long road across Texas a lot of horse-trading had gone on as the expedition was always in need of fresh mounts. By the time they crossed into Mexico the horses the Confederates rode, as well as those in their remuda, bore a variety of brands. Most of the horses had been acquired in honest trade, but many were no doubt prizes of the successful battles the expedition had fought on the way south.

While the negotiations were in progress between Governor Biesca and Shelby, the members of the brigade came and went in town as they pleased. The pause in Piedras Negras was a welcome respite for the men after the long exhausting crossing of Texas and they took full advantage of the diversions the town offered. They roamed the streets, drank at the local saloons, lounged in the plaza and competed for the attentions of the local senoritas.

All this aggravated some of Biesca's troops. Sinister looks passed between the two groups as they met each other in the streets. There was jealousy enough to go around as Mexican soldiers steamed at what they considered the unearned benefits these Gringos were reaping at their expense. Perhaps as an outgrowth of this resentment a rumor began to circulate that many of the Confederates were riding stolen horses, and stolen from whom seemed not to matter.

Though the men had the liberty of the town Shelby still demanded discipline, and drills and guard duty were seriously observed. Shortly after the completion of afternoon drill Ike Berry, a battle-hardened veteran of dozens of battles and long marches, sat languidly astride his horse in front of a local cantina. Ike wore a long yellow beard, and his hair hung to his shoulders giving him the appearance of a wild mountain man. Taking chance notice of Ike slumped in the saddle Biesca's men might have mistakenly assumed he was dozing and off guard.

A young Mexican captain stepped away from a small group of Juarista soldiers standing nearby and approached the grizzled Confederate horseman. Perhaps encouraged by drink or emboldened by the resentment he and his comrades felt for their Gringo visitors, the Mexican officer took hold of the reins of Ike's horse and broadly accused him and the other Americans of being horse thieves. He might yet have safely walked away, but then he grimly ordered Ike to dismount.

In a flash, hell's fury erupted. Suddenly, Ike stood in his stirrups as he simultaneously drew his saber in one motion. Then sweeping the deadly blade in a downward arc, he severed the Mexican's arm at the shoulder. It all happened so suddenly that there was a momentary silence as the stunned witnesses stared at the horrific sight of the victim's arm still grasping the reins of the horse as his body fell away; blood spouting from the stump with each heartbeat.

The other Mexican troops immediately fanned out and drew their weapons. An explosion of gunfire erupted through the street just as one of Shelby' troopers ran to inform their commander of what was happening. His report would be unnecessary.

Shelby heard the outbreak of gunfire and came running. The sight that met him was a familiar one and his instincts took over. Martin Kritzer, the brigade's bugler stood nearby. "Sound the rally! Sound the rally!" Shelby shouted as he mounted up and charged towards the gunfight. At the sound of the bugle the rebel troopers came pouring out of local cafes, stores and cantinas believing that Biesca had double-crossed them.[2]

The Missourians went into action without orders. One group of rebels immediately charged down to the river bank to secure all the boats while D.A. Williams led a small party of men in scattering the Mexican troops who were guarding the cannon. The Confederates began to man the guns as the Mexicans fled.

By the time Shelby and Biesca arrived on the scene a full-scale battle was in the making. Both men immediately began trying to defuse the situation and ordered their men to cease-fire. Their quick and resolute action calmed the combatants and prevented the battle from spinning completely out of control. The fighting eventually stopped, but seventeen Mexicans and two Americans lay dead.

That night a strange quiet settled over the town and both sides slept uneasily. A blanket of mistrust had fallen across the town. Biesca worried that Shelby might decide to keep his re-captured cannon and even perhaps turn the Napoleon Howitzers upon him. However, diplomacy reasserted itself and Shelby and Biesca were able to negotiate a precarious truce that held until the Americans left Piedras Negras. In the end Shelby returned the guns and Biesca handed over the money in payment as agreed.

Before breaking camp at Piedras Negras another unfortunate incident occurred concerning some of the horses that Shelby's men rode and their respective brands.

As soon as possible preparations were made to march and a pack train was formed. At daybreak Shelby's Expedition headed southwest toward the Sabinas River. As they approached the small village of Fuente a lone horseman rode up to join the column with a warning for the Americans of an impending ambush ahead. Shelby's instincts told him that the warning was real and realizing that the Mexican rider who had delivered it was now vulnerable, Shelby asked if he would like to join his brigade for the protection he could offer him.

"No Senor…this is evil country," he said. "Beware of the Sabinas." Without another word he galloped away.[3]

The odd encounter and the rider's warning spooked Shelby a little, but he had no reason to ignore it. He passed the order to his men to remain on high alert. To the south lay Monterey, a hundred and sixty miles ahead where the Great Plateau of Mexico commenced.

On they rode, past Fuente and through the town of Gigedo on the eastern edge of the Sierra Madres. Heat, dust and dehydration were constant irritants, unseen enemies that sapped their strength and detracted from the desolate beauty of arroyos, canyons and chaparral through which they passed.

A day's ride short of the Sabinas River the expedition reached a small village, the name of which went unrecorded. The approaches to the town seemed peaceful enough, perhaps too peaceful. To ears attuned to years of combat the silence was unnerving.

They had hardly entered the streets of the town when the cause of their anxiety became evident. No life greeted them; not a single living being, not even a dog. The place was a massacre site. Bodies' lay everywhere in grotesque positions. Some had been hung up and skinned alive, "with huge strips of skin dangling from them as tattered garments might hang." The broken and torn corpses of men and women were scattered throughout the town. Most shocking of all were the murdered and mutilated bodies of small children. Some people had been roasted alive over slow burning fires, the ashes still smoldering. Wild animals and vultures had been at their work devouring the corpses lying unprotected under the open sky.

Four years of war had not prepared the Confederates for what they saw in this little Mexican town. They were all battle-hardened soldiers and no strangers to blood and death. They had seen war in all its horror, but this kind of carnage was new to them. Some of the men were physically sickened.

It was clear to Shelby that the village was beyond help, so he ordered the brigade to a full gallop to put the scene behind them as quickly as possible. Taking time to bury dozens of mutilated and burnt bodies would have exposed the men to ambush, and the psychological damage to the troops would have been devastating; so they rode on.

Shelby's orders were not those of a man devoid of sympathy or of civilized compassion. He and all the men were moved to emotions of pity and vengeance, and no doubt they all were beginning to realize that they had placed themselves in dangerous straits. Obviously, no one was in control of Mexico; not the French, not the Mexican Army. No one.

With the scene of the massacred village fresh in his mind Shelby's battle-honed instincts were aroused. The warning he had been given earlier must have come to mind: "This is evil country…Beware of the Sabinas." Whatever band of ruthless cutthroats responsible for the butchery they had just seen was probably aware of their coming and a river crossing would be a perfect setting for an ambush. Having seen their handiwork Shelby knew there would be no quarter given if they had to fight them.

Shelby said to Ben Elliot, "If worst comes to worst, Colonel, keep one pistol ball for yourself."[4]

The perpetrators of the massacre were most likely Lipan Apache Indians and Mexican bandits riding together; an evil, murderous company of marauders. Shelby approached the Sabinas cautiously. First he sent D.A. Williams at the head of twenty men to cross the river and scout ahead. He also threw out two hundred troops to ride as a vanguard several hundred yards to the front of the main column. Their orders were to charge ahead and cross the river if they should come under fire and secure the furthest bank and cover the crossing of the rest of the brigade.

The approach to the river was along a rocky road that angled down toward the tree-lined banks of the stream. A waterfall's roar could be heard in the near distance downriver. The troops of the forward element were just descending the road to the crossing when a volley of gunfire was heard from across the Sabinas where D.A. Williams' scouting company had come under attack.

At the sound of firing the two hundred men in the vanguard charged forward and into the river at a gallop. Shelby was right behind them with the main body of the brigade and gained the far bank while the rest of the men were still crossing the river. There he found Captain Williams who had fallen back to report to Shelby that the guerilla force had been encountered just a quarter of a mile beyond the crossing. He estimated their strength at approximately eight hundred men.

A group of renegades had hidden themselves in the trees and brush along the riverbank, waiting until the main Confederate force was in the process of fording the river. As Shelby and Williams conferred, the ambush was sprung and the concealed gunmen opened a murderous fire upon the men crossing the river. Several saddles were emptied before the entire force could make it across the strong current of the Sabinas and begin to reform for an attack.

The bodies of the dead Americans swept downstream and over the falls as a bugle call sent Shelby's brigade into battle formation. Colonel Elliot was ordered to break off and deal with the renegade butchers who were firing from downstream, as the rest of the old Iron Brigade prepared to give their new adversaries a taste of their own medicine.

As Colonel Elliot charged the enemy along the bank of the river, Shelby's main force thundered toward the renegade camp. The vengeance in their hearts was almost palpable as they all remembered the massacred women and children of the village.

The Indian and Mexican guerillas attempted to form up to meet the hurricane coming at them, and although the numbers were fairly even, they really had no chance. The guerillas were more like an unorganized gang; deadly against the defenseless, but this day they were facing a veteran military force of disciplined, battle-tested killers. And they were terribly motivated. The rebel battle cry echoed above the din of gunfire and thundering hooves as the Missouri horsemen maneuvered into two lines just before striking the enemy like a runaway locomotive, riding right over and through them.

Then wheeling in unison they charged back through the hapless guerillas again cutting bloody swaths of destruction like a scythe through a wheat field. A pall of black powder smoke and dust clouded the battlefield as the hand-to-hand fight became intimate and personal. Soon the renegades were running for their lives in all directions. Shelby's men rode them down showing no quarter as sabers rose and fell and pistols roared. Some broke for the river and tried to swim for their lives. The rebels easily picked them off, cursing profanely as they fired and fired again until the river was full of floating bodies drifting towards the waterfall.

In less than an hour the bloody work was over. More than four hundred enemy corpses littered the battlefield, not counting those who had died at the river. There would be no Christian burials for the fallen enemy on this field of battle. During the short but costly fight, Shelby's command may have killed more than five hundred of the enemy. In doing so, they also sent a message to other gangs of assassins and cutthroats in the territory: a new aggressive force had entered Mexico, and it was called the Missouri Cavalry Division.

Shelby lost twenty-seven men with over thirty seriously wounded. The dead were wrapped in blankets and buried with military honors and a salute from a volley of carbines. Similar scenes would be re-enacted many times before the rebels reached Mexico City.

They remained in camp near the Sabinas for almost a week as the wounded were treated. Regular patrols scouted and kept up a continuing vigil as part of Shelby's determination to maintain military discipline and battle readiness. He realized that this was their only hope of surviving the march through the gauntlet that was Mexico in those years following the American Civil War.

Near the end of the first week of July the brigade broke camp and began to climb into the foothills of the Sierra Madre Oriental moving towards Lampazos.

The Sierra Madre Oriental is one of two mountain ranges that parallel each other and extend southward from the American border to the Valley of Mexico. Its average width is two hundred miles and its highest peak, Cerro Peoa Nevada, stabs 12,000 feet into the sky. The Oriental comes progressively closer to the coastline as it approaches Mexico City.

Connected to the Sierra Madre Oriental by several volcanoes in the south is the other mountain range, the Sierra Madre Occidental, which parallels the Gulf of California. It runs southward following the Gulf of Cortez. Depending on elevation, the terrain making up these two mountain ranges can vary from low lying peaks and ridges covered with pine forests to sub-tropical wilderness.

Shelby's Expedition was traveling along the eastern side of the Sierra Madre Oriental. As they climbed higher, fatigue became a constant companion that required more frequent stops to rest both men and animals. The physical stress of crossing the mountains added to their misery and interfered with their timing. They had reached this region at the height of the rainy season and almost every afternoon the rains came in torrents. At night the temperature dropped dramatically making camp comfort nearly impossible.

Having passed the high mountains the terrain leveled off into a small plateau across which the road to Lampazos meandered. All the while as they moved along guerilla bands that stayed out of rifle range shadowed them. Constant vigilance was necessary, especially as the brigade camped in the evenings because of snipers who would creep in close and try to get off a lucky shot or two. They also had to post pickets to prevent night raids on their horses.

Eventually they moved into a little valley that resembled an oasis with palm trees sparsely lining a mountain stream. Grass was abundant in the immediate area so they made camp at this place known locally as Paso de Palmas.

After most of the men had drifted off to sleep that night a Cuban exile turned outlaw named Antonio Flores led a hundred of his bandits to the edge of the camp. They had tethered their horses on the other side of the mountain from the rebel camp and slipped by the sentries. Their plan was to creep up on the sleeping Americans and spring an ambush. A heavy, low-lying mist coming off the river reduced visibility to only a few feet, which gave the bandits a distinct advantage for their plan.

As the bandits were still getting into position on the slope of a ridge running along the riverbank, one of Shelby's pickets realized that the enemy had penetrated their defenses and opened fire without warning. The startled raiders returned fire at close range, but the camp had already been roused. Groggy and half dressed, Shelby's men grabbed their pistols, sabers and Bowie knives and began running towards the enemy position.

Unable to see through the mist more than a few yards in front of them, the Confederates advanced and began to fire toward the bandits' muzzle flashes and at shadowy specters in the dark as they advanced. Considering the murky atmosphere it was provident that none of the rebels killed one another in the darkness. They could only recognize each other by their voices, so they kept up a constant communication by loudly shouting commands and directions in an attempt to stay organized. Soon enough they closed with their foes and cut them to pieces. Among those who fell was their leader, Antonio Flores.[5]

The bandit force fared badly in the battle, losing approximately seventy killed and seventeen badly wounded. The Confederates were extremely lucky that night. Their casualties were eleven dead and six wounded. It could have turned out much worse.

Another funeral ceremony and rifle volley for fallen comrades, and the brigade set out just after daybreak along the rock-strewn road to Lampazos. Along the way they passed the sheer walls of the towering Mesa de los Cartujanos atop which Benedictine fathers had built a mission a hundred years earlier. The village of Lampazos sat at the foot of the San Juan Range about half way between Peidras Negras and Monterey.

Monterey, "The City of Saddles," was held by General Pierre Jeannigros who commanded five thousand French and Mexican troops, among them four regiments of the legendary Foreign Legion which included a multitude of nationalities: Americans, Englishmen, Arabs, Turks, Swiss and Belgians.

Jeannigros and his troops were Maximilian's front line in the war against the Juaristas, though in reality the French never actually held more than a few towns in Northern Mexico, and even those were tentatively controlled. Even with over 40,000 troops in Mexico the French were still in a precarious position and Jeannigros' command was especially exposed in the wild northern territory. The Juaristas, and the outlaws who thrived in such uncertain times, on the other hand, were always a threat and roamed the countryside at will.

Jeannigros had a mindset and a record for the kind of work often employed to control recalcitrant populations. He was a thirty-year career soldier who had earned a bloody reputation as a result of his brutal reprisals against Mexican peasants and guerilla prisoners alike. As second-in-command he had chosen a man of like temper, sixty-year old French Colonel Francois Achille Dupin. Described by his enemies as a sadist Dupin was known as "The Tiger of the Tropics."

Bronzed by the sun and in excellent physical condition for a man his age, Dupin stood six feet tall and was a veteran of over fifty military campaigns. In China during the Peking Campaign he led an assault on the Emperor's palace, which earned him a reputation as a berserker and looter. His actions also led to a court martial that stripped him of his rank and decorations and sent him into exile in Mexico where he was eventually reinstated with a quasi-military command in the Contra-Guerrilla force under Jeannigros.

In Mexico he earned his newest nickname by employing a barbaric tactic he called "the economy of amputation."[6] "When you kill a Mexican, that is the end of him," he said. "When you cut off an arm or a leg, that throws him upon the charity of his friends, and then two or three must support him. Those who make corn cannot make soldiers. It is the economy of amputation."[7]

These were the kind of men who awaited the arrival of Shelby's Brigade, and they had no reason to welcome them warmly. In fact, Jeannigros had become enraged upon hearing of Shelby's sale of his guns to Governor Biesca at Piedras Negras. He coolly promised to capture and hang every man in Shelby's command and Dupin wanted to charge out and engage the Confederates in battle. Nevertheless, Jeannigros decided to wait. Besides, he had a healthy respect for the reputation of Confederate Cavalry and was unwilling to make the first move even though he commanded over five thousand troops and outnumbered the Americans five to one.

Shelby was in a fighting mood when the expedition approached Monterey in late July 1865. The extreme ordeals faced by Shelby and his troops on the march thus far had left them all anxious and out of sorts. Besides, he had never liked the idea of joining the French to begin with. Four miles from the French garrison Shelby drew his men up in a long line of battle and had his adjutant, Major Edwards, compose a message to Jeanningros. Translated into flawless French by Governor Reynolds and carried into Monterey by John Trailkill and Rainey McKinney under a flag of truce, the note read:

> General:
>
> I have the honor to report that I am within one mile of your fortifications with my command. Preferring exile to surrender, I have left my own country to seek service in that held by His Majesty, the Emperor Maximilian. Shall it be peace or war between us? If the former, and with your permission, I shall enter your lines at once, claiming at your hands that courtesy due from one soldier to another. If the latter, I propose to attack you immediately.
>
> Very, respectfully,
> Yours,
> J.O. Shelby.[8]

By all accounts and witnesses, Shelby seemed not to care which path the Frenchman chose. He may have actually believed his small brigade could whip the French mercenary force if need be, or he may have been employing the tactic of psychological warfare in an attempt to bluff Jeannigros. But, why would his message hint at a level of mistrust of the French? Either way, it seems strange that Shelby would announce his arrival in such a challenging manner, especially considering that his own men had chosen to join these very French troops.

As events were later to prove, Shelby's instinct to be wary in approaching Jeanningros was correct. After all, there were no guarantees that Maximilian would accept the Confederates' military services. And though Shelby had no way of knowing that Jeanningros had threatened to execute the Confederates, his assertive posture toward the French general was masterful. It seemed to arouse in Jeanningros a sense of respect for a comrade-in-arms, or perhaps he simply admired Shelby's bold letter, because he proved to be a superb host treating the Confederates "with the courtesy due from one soldier to another."

The festivities continued for several days while Shelby and Jeannigros engaged in meetings, dinners and celebrations. Jeannigros did not have a high opinion of Maximilian and told Shelby as much. Both of the veteran soldiers were in agreement on two points: Maximilian would not be able to conquer Mexico without support, and that the United States would soon feel compelled to intervene, thus threatening a war that France did not want.

Shelby provided his own analysis of the situation for Jeannigros:

> "I don't know what you think of things here…or what the future has in store for the Empire, but one thing is certain and I shall tell you the plain truth. The Federal government has no love for you French or your French occupation of Mexico. If diplomacy can't get you out, infantry divisions will. I left a large army concentrating on the banks of the Rio Grande and all the faces of the men were looking straightforward into Mexico. Will France fight? I think not; if your Emperor had meant to be serious in this thing he would have formed an alliance with Jefferson Davis long ago. Now it is too late. Nevertheless, if you will give me a port as a basis of operations, I can organize an American force capable of keeping Maximilian on his throne, composed of men who would be delighted to fight Yankees again. If left discretionary with me, that port shall be either Gauymas or Mazatlan, on the Pacific coast. The Californians love adventure, and many leaders among them have already sent me messengers with overtures. I am not informed as to the wishes of the Emperor; but the French cannot remain and he cannot rule over Mexicans with Mexicans. The plan I propose is his only hope. You

> know Bazaine better than I do. What would Bazaine say to all this?"[9]

Marshal Francois Achille Bazaine had recently become the overall military commander of the French expeditionary force in Mexico. Most of his illustrious career had been spent with the French Foreign Legion, but his reputation was honored throughout the French military services. In Spain he had led the Foreign Legion against the Carlists in 1837, receiving the Legion d'honneur for gallantry in combat. He spent nearly half of the next twenty years fighting in North Africa where he rose to brigadier-general. Bazaine earned further military honors in the Crimean War, where he also met future Confederate General John Magruder who was serving as an observer for the United States military.

Bazaine arrived in Mexico in 1863, and rejuvenated the French military campaign, driving Benito Juarez into near exile. But while his military credentials were impeccable and his military rank made him the real power in Mexico, he was at the time struggling in his relations with Emperor Maximilian, an association that would continue to deteriorate. He was a battle-hardened soldier and Maximilian was a politician who understood little of the military realities in Mexico.

Because of his unsteady relations with Maximilian, Bazaine was something of a wild card when it came to the question of how to make use of the former Confederates flooding into Mexico. And although he appeared to be taking a hard line when dealing with the Confederates he nevertheless seems to have given careful consideration to the idea of an alliance with the rebels.

This was despite warnings from the Belgian legation in Mexico that employing the rebels as a part of the French Army would pose a danger to Maximilian's Empire by inviting intervention by the United States. The Belgian assessment explained that at present the North was gradually disarming and disbanding its troops, and if matters were quietly allowed to take their course the danger would pass in six months.

If, on the other hand, Grant felt compelled to send a large army into the southwestern frontier, not to mention Mexico herself, disastrous complications might result for France. The Belgian counsel advised that if Jefferson Davis [or remnants of his army] crossed the Rio Grande they should be shown consideration as soldiers and gentlemen, but given every encouragement to proceed immediately to Europe. Otherwise, their presence in Mexico would unfortunately, be only a provocation for the United States to invade Mexico.[10]

Jeannigros believed that Bazaine would be receptive to Shelby's offer. He himself agreed with Shelby's assessment of the situation in Mexico and gave him permission to march to the Pacific and set up his base of operations at whichever port the Confederate commander chose for recruiting his new army for the Empire. The final decision, however, would be Maximilian's; and the emperor was no soldier.

When the celebrations and conferences ended it was decided that Shelby would march to Mazatlan by way of Saltillo, past the old battlefield at Buena Vista. It was there during the Mexican-American War that Colonel Jefferson Davis commanded a regiment of Zachary Taylor's army that included both southerners and northerners against the Mexican Army. General Jeannigros and Shelby shook hands amiably and Shelby led his brigade into the sub-tropical climate of Mexico, leaving behind Monterey and its orange and lemon groves. Their route took them through Saltillo along the high road to Parras, and as they climbed, the air became thinner and cooler.

Two days later, before they reached Parras, Shelby's camp was attacked during a driving rainstorm by Mexican guerillas. Only strict discipline and adherence to military procedures in enemy territory saved the expedition. As always, Shelby had posted several pickets around the camp that alerted the men just in time.

In what turned out to be one of the most desperate fights of the march, some four hundred Mexican renegades simultaneously attacked from both the front and rear as the thunderstorm raged. The Iron Brigade was in a foreign wilderness with no real allies and no means of support or reinforcement. Their survival depended solely on themselves and they fought for their lives with a desperation that only men who have faced death in such circumstances could truly understand.

There is no record of Shelby's losses during the ruthless night attack, other than that they were reported to have been heavy. Since crossing the Rio Grande the brigade had lost over fifty men and had killed nearly one thousand of the guerillas and bandits they had fought along the way. The Civil War may have been over for the rest of America, but not for these men. For them it must have seemed as if they were living in a universe of battle. The next morning more graves were dug, prayers said, and another volley fired before the expedition mounted up and continued on to Parras.

Colonel Maguerite Jacques Vincent DuPreuil, another veteran officer of Louis Napoleon's Army, commanded the French garrison at Parras and would become a serious obstacle to the plans Shelby and Jeanningros had laid in Monterey. Several other Confederates were already in Parras. With them was Colonel Davis S. Terry, the former Chief Justice of the California Supreme Court.

Word of the arrangement between Shelby and Jeanningros had already reached Parras and had been the topic of much discussion among the Confederates waiting there. Shelby was quickly informed that Colonel DuPreuil had recently received a dispatch from Marshal Bazaine suggesting that Shelby's brigade was the spearhead for a possible American invasion. Forewarned, DuPreuil was in a confrontational mood when Shelby and the brigade arrived and was determined to challenge the expedition's plan to march to the western coast of Mexico to raise a volunteer army of Americans.

The situation was aggravated and made even more complicated by unreliable intelligence reports, which prompted Marshal Bazaine to send his misguided warning to Colonel DuPreuil. When Shelby sought an audience with the French commandant he found him intoxicated and rude, even refusing to take Shelby's extended hand. Acting upon Bazaine's information the petulant and explosive DuPreuil wasted no time in informing Shelby that the brigade would not be allowed to continue their march to the Pacific to raise an army. Instead Shelby must divert directly to Mexico City.

This was not what Shelby expected to hear. After all, he had arrived in Parras with the blessing of Jeanningros, a subordinate of

Marshal Bazaine himself. Shelby hardly had time to respond to this unexpected demand and the manner in which it was delivered. Neither did DuPreuil pause for a response, but immediately accused the American of coming to Mexico only for plunder, and he *ordered* him to obey Bazaine's new directive.

Shelby rose from his seat, put on his hat and said slowly and deliberately, "I am mistaken it seems; I imagined that when an American soldier called upon a French soldier, he was at least visiting a gentleman. One can not always keep his hands clean, and I wash mine of you because you are a slanderer and a coward."[11]

Where DuPreuil came from such a rebuke was taken as seriously as in Shelby's South. He seized the hilt of his saber and began to draw the blade from its scabbard. At the same time Shelby's hand flashed to his pistol. The two old soldiers' actions were clear; death was only a whisper away and only blood would wash away the insults.

Both men glared at one another as the other French officers in the room came to their feet, their weapons at the ready. Through clenched teeth, DuPreuil ordered Shelby to remove his hat. Shelby's blood was up and he fiercely replied to DuPreuil's command, "Only to beauty and to God; to a coward, never!"

The stare down continued for a few alarming moments before DuPreuil broke the silence. "Retire! Retire immediately lest I outrage hospitality and dishonor you in my own house. You shall pay for this! You shall apologize for this."

Shelby made a move for the door in an act of finality. Then a pause, a look of defiance in DuPreuil's direction, and with emphasis he said, "Never."[12]

No matter what Shelby might have thought about DuPreuil, the French officer was no coward; and Shelby's insult had been spoken in the presence of the Frenchman's subordinates. His inflexible sense of honor had been questioned, and he could not let the insult stand. Accordingly, a duel between the two antagonists was scheduled for the following morning. With that, all the Confederates' future plans and possibilities seemed to evaporate. All that they had risked, all that they had endured, hung in the balance of implacable honor. Even if Shelby won the duel, what would be their fate should he kill DuPreuil?

DuPreuil, as the challenged party, had his choice of weapons. Known as an expert swordsman, he naturally selected the saber, but when he learned that Shelby had a fresh gunshot wound near his right elbow, he graciously agreed to use pistols instead. DuPreuil's chivalrous gesture might have been a serious mistake because Shelby was a deadly pistol shot and would probably have killed him outright.

Fortunately, the burning fuse was extinguished when Colonel Jeannigros unexpectedly arrived later that night. Jeannigros had recently been promoted to brigadier general and was already wearing the epaulets of his new rank. When he heard of DuPreuil's drunken behavior with Shelby, he placed him under arrest. Jeannigros called another meeting the next morning during which DuPreuil apologized to his American guest. The duel, and almost certain tragedy, was averted.

Though DuPreuil and Shelby's honor had been satisfied for the time being, Bazaine's order was not altered. Shelby and his men would not continue to the western coast and Mazatlan; they would march to Mexico City to meet with the Emperor.

The brigade moved out southward towards Queretaro and La Encatada where the altitude was over six thousand feet. San Luis Potosi lay two hundred miles beyond. In their path stood rugged desert trails, mountain passes, and tropical valleys rich with exotic plants and fruit trees. From San Luis Potosi it was another four hundred miles to Mexico City.

Approaching Matehuala by way of the narrow Carnero Pass, the men marveled at bare sierras on both sides. Within two miles of Matehuala they could hear the sounds of battle echoing through the mountain passes. The heavy booming of artillery and the crack of small arms fire told the veteran soldiers all they needed to know; a substantial fight was in progress at the French garrison posted in the small town just ahead. They surmised that the French were under attack and that the Juaristas were between them and Matehuala with their backs to the brigade.

Shelby waited until nightfall to move his brigade forward toward the rear of the Juarista battle lines drawn tightly around the beleaguered French compound. Major Henri Pierron of the French Army held the town with five hundred troops of the 82nd Infantry, and Shelby could see plainly that they were in serious straits. Two thousand Mexican rebels commanded by General Mariano Escobedo surrounded Matehuala and the French works. An assault was certain to come with the next sunrise and the chances were good that the French would be overrun.

James Cundiff and Elias Hodge volunteered to carry a message of hope through the Mexican lines to Pierron. After some close calls they finally reached Major Pierron and explained to him that the brigade was in position just beyond the Mexican lines and that Shelby intended to attack their rear at dawn. At the sound of their gunfire Pierron and his men were to simultaneously charge the front of the Juarista positions. Pierron readily agreed to Shelby's plan.

At sunrise the Mexican siege guns began their bombardment of the French garrison. Shelby's brigade was already formed and prepared to charge. Coming forward in companies by columns of fours they first advanced to a trot, and then to a canter as they crossed an open field. They had no way of knowing at the time, but this would be the last charge of the Iron Brigade.

Shelby focused his attack directly on the Mexican siege batteries. Rising in his stirrups he pointed toward the enemy guns and shouted, "Gallop, Ho!" Above the thundering tumult of the charging horses the rebel yell could be heard distinctly.

Major Pierron could not mistake the signal and ordered his men to sally forth. The Mexicans were not sure at first if the Confederate force advancing from their rear was friend or foe. Within seconds the answer came crashing into their ranks. Shelby's brigade hit them and swept everything before them like a flood. The confusion inflicted by the first charge had achieved its purpose. The dazed Juarista soldiers wavered as the brigade wheeled for another charge. Their pistols, however, had done their work on the first pass, and the Mexican lines began to disintegrate. Within a few moments the Juarista's were fleeing for their lives in all directions while the French Cuirassiers dashed in from their flanks with sabers drawn to finish the bloody work.

When it was done, and the garrison rescued, Major Pierron escorted Shelby and his men into the town where they would be accorded the proper honors they deserved. The festivities lasted for several days during which John Newman Edwards wrote, "Pierron turned Mahtehuala into a paradise. There were days of feasting and mirth and minstrelsy, and in the balm of the fragrant nights the men dallied with the women."

Following the healing rest and relaxation, Shelby's Brigade took their leave of the grateful French and headed towards San Luis Potosi. At one point they found themselves riding southward under a full, tropical moon through fields of purple and pink blossoms and were engulfed by their perfume. The poetic nature of John Newman Edwards could not be contained. "The whole earth smelt sweet," he wrote.[13]

General Felix Douay, who was a confidant of Marshal Bazaine, commanded the French garrison at San Luis Potosi. And although General Douay welcomed Shelby with offers of an immediate command in the French Army, Bazaine had other ideas. Political intrigue and the shifting sands of fate seemed to follow Shelby's Expedition like the shadow of doom. Aside from selling the artillery pieces to Governor Biesca at the border the Confederates had done nothing to give Maximilian and his French protectors reason to dread them. Although he and his men had performed brilliantly in rescuing the French garrison at Matehuala, their successful action would come under suspicion within a few days and even work against them.

For the moment, however, they were being received, at least outwardly, as heroes. Their reputation had preceded the Confederate horsemen's arrival at San Louis Potosi. General Douay not only offered Shelby a command in the French Army, but he also charged him with a special mission: find and kill the Mexican renegade, General Luis Figueroa.

Figueroa was a merciless killer who had earlier lost an eye to a bullet from an American's pistol. For that he hated all Americans. The French were his mortal enemies because of their presence in his homeland and he now saw the Americans as no better than the French. It was but a small step from there to a hatred of anyone from outside his home country, and he was unrelenting in his cruelty towards all foreigners.

One who harbors blind hatreds is often beyond redemption. Even his own people were not safe from his atrocities. Many Mexican merchants, farmers and peasants paid tribute rather than face his retribution. Figueroa would become a scourge to the Americans who would later take up land in Mexico and attempt to establish an agricultural settlement between Vera Cruz and Mexico City. They would call it Carlota in honor of the empress.

The specter of Figueroa haunted Mexican and foreigner alike. Three separate French columns had been sent out to kill or capture him. As John Newman Edwards described the campaigns, they "had been sent into the valley of the shadow of death and had returned, those that were left of them soldiers no longer but skeletons whose uniforms served only to make the contrast ghastly."[14]

The Confederate brigade enjoyed the hospitality of General Douay for the next few days as they awaited Marshal Bazaine's decision. When it finally came the message was brusque and to the point: "Bid the Americans march immediately to Mexico City."

Bazaine had never fully trusted the Confederates' motives, and he also believed that the time had passed to recruit an army of these North Americans for service with Maximilian. He also believed the news of Confederate cavalry fighting alongside the French against the Juaristas in the battle for Matehuala would alarm the American government in Washington. This alone, he reasoned, might provide Grant and Sheridan with the final motivation to expedite military action against Maximilian. Bazaine had received a number of communications, dubious though they might have been that the United States was contemplating such action to force Maximilian and Louis Napoleon out of Mexico. In Bazaine's mind the Confederates were more of a liability than an asset. There would be no appeal.

Though he was uncertain of the Confederates he nevertheless authorized the French garrison at Matehuala to re-supply them for their march to Mexico City. There Emperor Maximilian, the final judge of their future fate in Mexico, awaited them.

Perhaps the men sensed that their expedition was running out of options for it was during this part of the journey that discipline within the brigade began to break down. Some of the men became surly and questioned the orders of their officers. Rumors of insubordination began to circulate along with mutterings that strict adherence to military procedure and maintaining battle readiness was no longer required.

Then one evening the first hint of outright mutiny was seen when a few men refused to pull guard duty. Word that some troops were disregarding the orders of their officers, quickly reached the former brigadier general of the Missouri Cavalry Division. A serious breach of discipline had occurred and there was no time to let the chain of command solve the problem. Shelby approached the potential mutineers personally, and in a short, direct speech he told them that he cared not for apologies, excuses, or reasons for their behavior. To Shelby's mind their choice was a simple one: all would be forgotten if the men returned to their duties. He would entertain no questions. "Not another word," he told them, "but if within fifteen seconds by the watch you are not in line for duty, you shall be shot like the meanest dog in all the Empire."[15]

In less than the fifteen seconds allowed the would-be mutineers were in formation and ready to do their turn at guard duty. The leader of the Iron Brigade stilled ruled with an iron fist.

Shelby had promised to lead this expedition to Mexico City come hell or high water, even if it meant shooting any malcontents who threatened the cohesiveness and discipline of the brigade. His intervention and demand for obedience left no doubt that he meant what he said. They were all part of an organized military unit and they had better damned well act like it. After that things fell back into line.

There seemed to be urgency now in Shelby's pace. On they rode through Dolores Hidalgo and into the Valley of Laja, steadily descending from the highlands of the Sierra Madre into lush green fields where herds of cattle and horses grazed. Soon in the distance appeared the spires of San Miguel de Allende, the picturesque 16th Century city sat at the foot of Carro Moctezuma. Shelby's brigade was greeted by Colonel Miguel Lopez on the outskirts of town and was escorted over old-world cobblestone streets to their quarters.

Early the next morning the brigade headed out towards Queretaro, an ancient Otomi Indian township. There Shelby and his men made final preparations for their arrival at the ancient Imperial City.

After weeks of difficult and dangerous marching they were only a hundred and twenty-five miles from their destination. They were now crossing the high plateau tablelands of Central Mexico and the land of the Aztecs. A week earlier Shelby had sent Major Edwards and John Magruder ahead to gather as much information as possible and to inform the French officials that the Expedition was soon to arrive. This news no doubt caused a measure of excitement and apprehension among the foreigners and Maximilian loyalists in Mexico City.

On the day the Confederates approached the city a full battalion of Chasseurs d'Afrique was dispatched by Marshal Bazaine to escort the Confederates to their temporary quarters in the barracks of the French garrison. Shelby and his officers were given rooms at the Hotel San Carlos where they were joined by Major Edwards and Henry Watkins Allen, former Governor of Louisiana.

Allen had made the journey to Mexico by sheer will power. He had been wounded at Shiloh in April 1862, shot through the cheek; and was later crippled when both legs were shattered by an artillery round at the Battle of Baton Rouge. Like the other Confederates arriving almost daily in Mexico City, he was looking for opportunity. He would later become publisher of the *Mexican Times*, an English language newspaper dedicated to the interests of the American exiles in Mexico.

Shelby was in Mexico City by order of Marshal Bazaine, but he was anxious to meet the Emperor and make his case for military posts for himself and his men in his service. The meeting between Shelby and Maximilian was set for the next afternoon at 2:00 p.m., Thursday August 16, 1865, in the Palacio Nacional. Also attending the meeting were Marshal Bazaine, Commodore Matthew Fontaine Maury, formerly of the United States Navy and most recently of the Confederate Navy Department; General John Magruder, and, of course, John Newman Edwards. Acting as interpreter at the meeting was Count de Noue, Chief of Bazaine's civil staff.[16]

For two months Shelby had been rehearsing what he would say to the Emperor Maximilian, and now he attempted to make the most of his opportunity. He explained in some detail his plan to recruit an army for service in the French Empire, and when he had finished, Maximilian merely bowed his head forward as if asking to hear more. Shelby continued in his straightforward manner:

> It is only a question of time, Your Majesty, before the French soldiers are withdrawn. Because

> the War between (sic) the States is at an end, and Mr. Seward will insist on the rigorous enforcement of the Monroe Doctrine. France does not desire a conflict with the United States. It would be neither popular nor profitable. I left behind me a million men in arms, not one of whom has yet been discharged from the service. The nation is sore over this occupation, and the presence of the French is a permanent menace. I hope Your Majesty will pardon me, but in order to speak the truth it is necessary to speak plainly. [The Emperor told Shelby to go on and the Missouri general laid out his program.]
>
> The matter whereof I have spoken to you is perfectly feasible. I have authority for saying that the American government would not be adverse to the enlistment of as many soldiers in your army as might wish to take service, and the number need be limited only by the exigencies of the Empire. Thrown upon your own resources, you would find no difficulty, I think, in establishing the most friendly relations with the United States. In order to put yourself in a position to do this, and in order to sustain yourself sufficiently long to consolidate your occupation of Mexico and make your government a strong one, I think it absolutely necessary that you should have a corps of foreign soldiers devoted to you personally, and reliable in any emergency. I have under my command at present about one thousand tried and experienced troops. All of them have seen much severe and actual service, and all of them are anxious to enlist in support of the Empire. With your permission, and authorized in your name to increase my forces, and in a few months all the promises given here to-day could be made good.[17]

Shelby paused respectfully to indicate that he had completed his assessment. The Emperor remained silent, and then he arose without speaking directly to Shelby, and dismissed himself from the presence of his visitors.

Throughout the meeting Bazaine had said nothing. One might have thought he was not really listening except for the occasional smile that crossed his face from time to time as Shelby delivered his expert analysis and proposal. And, although he had serious reservations about the Confederates' motives, he had none concerning Shelby's. The old warhorse was impressed by Shelby's dynamic presence and military mind. All of the Americans present, Shelby, Magruder, Maury and Edwards, believed unanimously that, by the end of the meeting, Bazaine "was strongly in favor of the employment of the Americans, and had the bargain been left to him, the bargain would have been made before the end of the interview."[18]

On his way to the exit, de Noue passed by Shelby, paused and said,

> "It's no use. The Emperor is firm on the point of diplomacy. He means to try negotiation …with the United States. He thinks Mr. Seward is favorably disposed toward him …at any rate, his majesty is an enthusiast, and you know that an enthusiast reasons ever from the heart instead of the head. He will not succeed. He does not understand the people over whom he rules, nor any of the dangers which beset him."[19]

Two hours later de Noue returned from the presence of the Emperor with the verdict.

"Its no use I say again, General, the Emperor will not give you employment."

"I knew it," Shelby replied.

"How?" asked de Noue

Shelby's answer was:

> "From his countenance. Not once could I bring the blood to his calm, benignant face. He has faith

> but no enthusiasm, and enthusiasm such as he needs would be but another name for audacity. I say to you in all frankness, Count de Noue, Maximilian will fail in his diplomacy."[20]

Shelby was in a dark mood as he rode back from the meeting to notify his men of Maximilian's decision. All of his plans had disintegrated, and he perhaps felt as though he should have been more forceful in negotiations with his troops concerning their decision to join the French.

> We are not wanted, and perhaps it is best so. Those who have fought as you have for a principle have nothing more to gain in a war for occupation or conquest. Our necessities are grievous it is true, and there is no work for us in the line of our profession; but to-day [sic] as upon the first day I took command of you, I stand ready to abide your decision in the matter of our destiny. If you say we shall march to the headquarters of Juarez, then we shall march, although all of you will bear me witness that at Piedras Negras I counseled immediate and earnest service in his government. You refused then as you will refuse today. Why? Because you are all Imperialists at heart just as I am, and because, poor simpletons, you imagined that France and the United States might come to blows at last.[21]

At this point Shelby checked himself. Suddenly realizing that he was lecturing them he turned to his old, ironic wit, "I find myself quoting Latin when I do not even understand Spanish. How many of you know enough Spanish to get you a Spanish wife with an acre of bread fruit, twenty-five tobacco plants and a handful of corn?"

His men responded with a loud round of laughter and applause.

All that remained now was for the former comrades to decide their course and go their separate ways. But first Shelby and Bazaine met at Tacubaya, a hamlet on the outskirts of the city, where the two old soldiers talked late into the night about their military adventures. The following morning Bazaine took over twenty thousand dollars from his military chest and offered every trooper in Shelby's brigade fifty dollars; a gift from one soldier to another. Thereafter, the Iron Brigade went out of existence and "every man went the way that suited him best."[22]

Several of the Confederate exiles acquired new positions in a variety of occupations in Mexico. Commodore Matthew Fountaine Maury was perhaps best situated for a prize appointment in the Imperial government. As a United States Naval officer in the 1850s he had been the first to chart the world's ocean currents. His book, *Physical Geography of the Sea*, made him internationally famous and gave him entrée into heady circles in both hemispheres. On a trip to Europe he had been introduced to Maximilian von Hapsburg, Prince of Austria, and Admiral of the Austrian Navy. The introduction led to a friendship, which bore fruit at Maury's hour of need in Mexico. He was appointed Imperial Commissioner of Immigration and thousands of acres of prime farmland was placed at his disposal for distribution to the foreigners entering Mexico.

General John Magruder was the new Imperial Commissioner of the land office. Governor Thomas C. Reynolds became head of the American and Mexican Emigrant Company, chartered by Maximilian. Henry Watkins Allen received a loan from Maximilian's Treasury and began his English language weekly in Mexico City, the *Mexican Times*, with John Newman Edwards as editorial assistant.

General Walker H. Stevens, the former chief engineer of General Robert E. Lee's staff, was appointed engineer of the Mexican Imperial Railway. Governor Reynolds also became superintendent of two short rail lines running out of Mexico City. General B.H. Lyon, of Kentucky, and Brigadier General John

McCausland, of Virginia received appointments as government surveyors. Others received permission to march to the Pacific Coast and board steamers bound for China, Japan, Brazil, Europe and California. Approximately fifty of Shelby's troopers joined the French Third Zouaves and other units in the Imperial Army in Sonora.

General Sterling Price selected a colonization site of 500,000 acres in the Cordoba Valley sixty-five miles from Vera Cruz, and three days by stagecoach from Mexico City. It was a large tract of uncultivated land that had originally belonged to the Church and was later confiscated by Benito Juarez who, in turn, lost it to Maximilian's control in 1863. The valley ranged in elevation from eight hundred to two thousand feet above sea level, and when yellow fever was not raging, or guerillas raiding from the mountains; Cordoba was another Garden of Eden. It was to this exotic and remote tropical location that Shelby, Isham Harris, Sterling Price, and approximately two hundred other Confederate volunteers, came and decided to make their last stand as colonists.

John Newman Edwards, ever the poet and philosopher, proved almost prophetic when he wrote, "To the superficial observer the conditions of the affairs in Mexico in the latter part of the year 1865 seemed most favorable, indeed, to the ultimate and successful establishment of the Empire."[23]

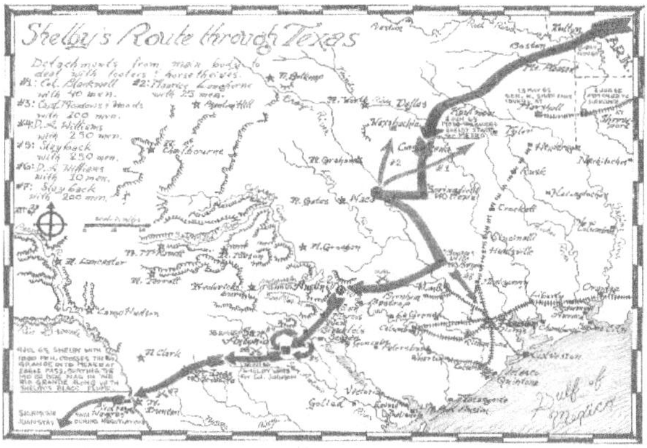

Shelby's Route Through Texas

Route taken by Shelby into Mexico

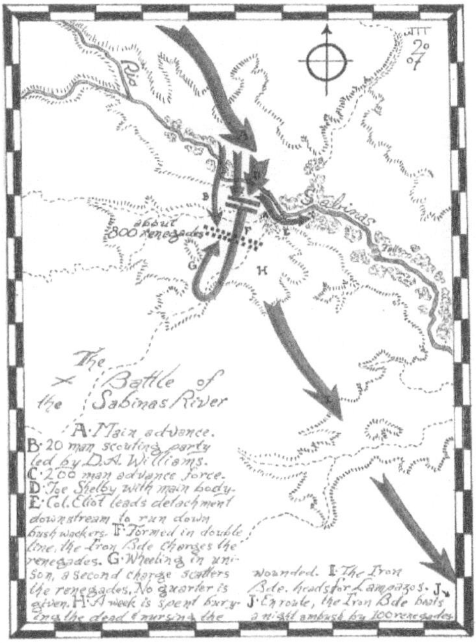

The Battle of the Sabinas River

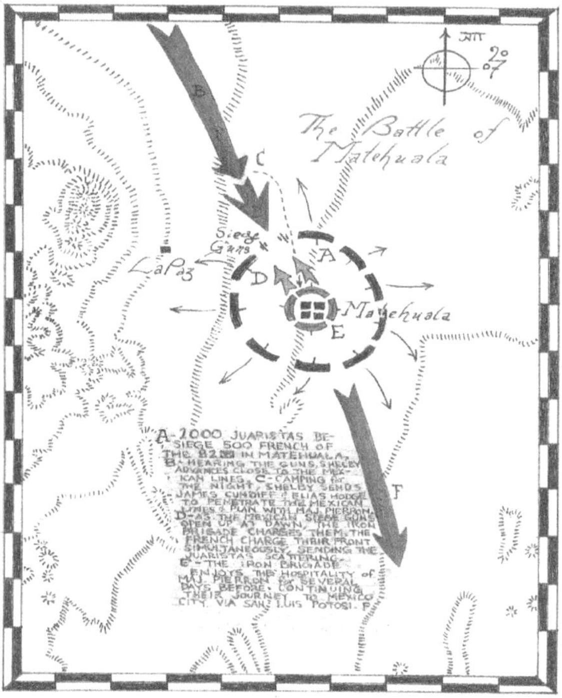

The Battle of Matehuala

The Empress Carlotta

Maximilian I and his Empress

Joseph Orville Shelby after his return to the United States

CHAPTER FIVE

Conception of the Colonization Program

While Shelby's adventurous expedition to Mexico was playing itself out the large-scale movement of Confederate troops and supplies into Mexico surprised the Federal officials who had accepted Kirby Smith's surrender. General Phillip H. Sheridan, who had been sent into Texas with an army of 52,000 men to occupy the territory and establish a provisional military government, soon found himself calling for more men to patrol the Rio Grande in an effort to prevent such immigration. Sheridan was furious because of the Confederate movement into Mexico, and relayed to Grant every report of Confederate migration he received, no matter how fragmentary or confusing. In June he wrote to Grant that:

> Kirby Smith, Magruder, Shelby, Slaughter, Walker, and others of military rank have gone to Mexico. Everything on wheels… artillery, horses, mules have been run over into Mexico. Large and small bands of rebel soldiers and some citizens, amounting to about 2,000 have crossed the Rio Grande into Mexico, some allege the intentions of going to Sonora.[1]

Sheridan began to make a nuisance of himself with his almost daily requests for permission to pursue the Confederates and to retake the material that had fallen into Maximilian's hands. Finally Grant wrote to President Johnson relating the substance of Sheridan's reports and endorsing his request for orders to cross into Mexico.

> Rebels in arms have been allowed to take refuge on Mexican soil, protected by French bayonets. French soldiers have fired on our men from the south side of the river in aid of the rebellion. Officers acting under the authority of the would-be empire have received arms, munitions, and other public property from the States…Large organized and armed bodies of rebels have gone to Mexico to join the Imperialists.

Suggesting further that the permission Sheridan requested be given, Grant advised the President that, "neutrality will not remove the threat."[2]

Though orders never came authorizing Sheridan to cross the border, he availed himself of every opportunity to create difficulty for the refugees. Thinking that the Confederates might try to join

Juarez, he conferred on several occasions with Liberal representatives to convince them that if the Confederates were allowed to join their movement they would, when successful, "be able to control the new Government, in fact, take possession of it, and that the disaffected rebels from the South would flock to their standard in sufficient numbers to hold the government." Finally he was able to report that "the rebels who went into Mexico have been defeated in their calculations, and have been forced to join the losing side" [meaning Maximilian].[3]

Sheridan felt that he had forced the Confederates to join Maximilian through his own activities and was now convinced that he would be permitted to cross the border. To lend a sense of urgency to his request he forwarded to Grant every message, regardless of its authenticity, that reported the organization of Confederates into Imperial military units. Still no orders came for Sheridan to plunge into Mexico, and the frustrated interventionist had to content himself with securing the border against any further emigration.[4]

While Sheridan "demonstrated" along the Rio Grande, the first of the ex-Confederate officers were arriving in the Mexican capital. There they learned that they were not the only Southerners who had chosen refuge in Mexico. Among them were General John Magruder and Mathew Fontaine Maury, celebrated oceanographer, naval astronomer and scholar as well as former official in the Confederate Navy Department. Both were helpful in introducing the new arrivals to Imperial government officials as well as helping to establish themselves.

In the months that followed, hundreds of Southerners of every rank, many bringing their families, made their way to Mexico. Until Sheridan successfully closed the Rio Grande crossings they came through Monterey, where they rested before continuing south through Saltillo and San Luis Potosi to Mexico City. Others found it necessary or more desirable to come by sea. Leaving from Southern ports such as Galveston or New Orleans on the Gulf, or Charleston on the Atlantic, they made their way to Vera Cruz where on foot and by horseback they crossed the shorter distance to the capital.

Whether as Sheridan believed, his hostile demonstrations on the border influenced the Confederate role in the Empire is uncertain. However there can be little doubt that the official attitude of the United States Government toward Maximilian's enterprise caused the Emperor's reluctance to accept their military services. Maximilian wished to establish, through diplomatic channels, peace and friendship with the United States. Should he accede to any plan to incorporate Confederate troops into his military organization, such overtures to Washington would be doomed. Furthermore he was not willing to trust so large and complete an organization of veteran American soldiers as an integral part of his Empire. Maximilian would not give the ex-Confederate soldiers military employment.[5]

It mattered only to the future of the Empire that Maximilian would not accept the Confederate exiles into his army. There were other plans under consideration that would provide a home and familiar occupation to every Southerner who wanted to leave the United States. Matthew Fontaine Maury and others had impressed Maximilian with their ideas on immigration and colonization of the idle lands of Mexico, and were busily drawing up plans for the Emperor's consideration even as the first of the Confederate soldiers were arriving.

Immigration, however, was not a new idea in Mexico, even to Maximilian. From the beginning of European colonization in the sixteenth century the King of Spain had relied on entrepreneurs to bring settlers at their own expense to establish Spanish culture, language and religion in the New World. The Church and its mission system, though partially subsidized by the Crown, had done the same thing through much of the Spanish era. Later, after independence from Spain had been won, it was Mexico's plan to bring American settlers to Texas in the 1820s to settle, pacify, and hold the land for Mexico against hostile nations. Early in the movement of the Reform the possibility of foreign immigration had been raised by Constitutionalists, and was by no means absent from Juarez' vision of his country's future. A decree was issued on March 13, 1861, clarifying and favoring "the individual or company inclined toward undertaking settlement in Mexico." Juarez was staunchly supported in such a proposal by the principle French newspaper in Mexico City, *L'estafette*, whose editor, Charles de Barres, insisted that for a long time everyone had agreed that the prosperity of Mexico lay in immigration. Barres urged that the first consideration of such an immigration program be to attract "elements which will mix with her people instead of dominating them."[6]

Due to internal problems and the French intervention, the Juarez government did little to inaugurate any significant immigration program. The French, however, were convinced that not only the success of their project, but the future of Mexico was dependent upon immigration. They were fascinated by the success the United States had experienced with European settlers prior to 1860 and believed that by this means Mexico could be "regenerated" and North American domination of the country overcome. Those who were committed to immigration believed that the strife-torn nation could be pacified if foreigners were brought into the country in numbers too great to be assimilated by the native population. Such a program would also be beneficial in raising the level of social conditions, and of most importance to the French, would stabilize and advance her economy for the production of wealth.[7]

The example of the United States was held before the apprehensive Mexicans early in the Intervention to demonstrate what could be achieved in Mexico with significant numbers of European settlers under a liberal land policy. To further allay Mexican suspicions, the French revised the old fear of a menacing northern neighbor. European immigrants, the Mexicans were told, could be settled along the Rio Grande to discourage any incursions from the United States. The American Civil War, they reasoned, would not wear the "Yankees" down but would only increase their expansionist appetites.[8]

As the French encountered difficulties in persuading significant numbers of Europeans to immigrate to Mexico, North Americans, especially those of French descent in Canada and certain parts of the South, gradually gained favor. In 1863, *L'estafette* began to suggest the possibility of Confederate immigration, predicting that especially in the event of the downfall of the Confederacy, refugees would seek homes in Mexico. Barres countered charges by some Mexicans that such a migration would subvert their customs and change the entire character of the Mexican nation, with assurances that Southerners would present themselves as enemies only if rebuffed.[9]

Maximilian appointed a commission to further study the prospects of immigration from the Southern States shortly before the Civil War came to an end. The commission reported, "far from being a danger to Mexican nationality these adopted citizens would be a powerful support. The hatred they feel for the United States Government is a guarantee that they will never lend a hand to an enterprise against Mexico."[10]

These proposed projects for immigration were by no means secret information; therefore, it was only a matter of time before some entrepreneur with a predilection for wealth and power should appear on the scene. Such a man was William Gwin, an enigmatic character from California who had served a term in the United States Senate and had, through his many and varied activities, established a reputation as a promoter of the old school. In 1863, he went to Paris to persuade Napoleon to allow him to establish a colonization scheme in the states of Sonora, Sinaloa, Chihuahua, and Durango. Although the nature of his agreements with the French Emperor are still vague, it appears that he had obtained assurances from Napoleon that he would be supported in his scheme to develop the mining possibilities in the above mentioned states.[11]

Gwin's plans were delayed by the unsettled state of affairs in the areas of his concession and by Maximilian's unwillingness to cooperate with him. Gwin demanded that he be given the French troops which Napoleon had allegedly promised him to protect his operations, and Maximilian, reared in royalty and unaccustomed to enterprising demands, declined. Theirs was definitely a clash of personalities. Short of patience, Gwin went again to Paris in 1865. Again he was assured that he would be given "every facility and assistance" to permit him to begin inviting immigrants. Upon his return to Mexico he regarded himself as the unofficial agent of the French government for immigration. All that he needed was the signature of Maximilian, who not only refused but also ordered Gwin out of the country when he made a nuisance of himself.[12]

Although Gwin's plan was never executed, his activities created great excitement in North America. The United States Government as well as the Confederates believed that Gwin was fostering a colonization scheme that was wholly Confederate.[13] Napoleon had perhaps foreseen that some difficulty might arise with the United States and had instructed Gwin to abstain from all connection with the Civil War in the United States.[14] Whether Gwin had made any commitment to the Confederate government is unclear and the source of Richmond's understanding cannot be accounted for.

Before Gwin left Mexico there were others in the capital city with plans for settling Southern expatriates in the Empire. Most of these men, however, were not motivated by selfish considerations, but were refugees themselves who were seeking new homes. Matthew Fontaine Maury became the foremost figure in this movement, and was most instrumental in winning Maximilian's support because of a friendship dating back a number of years.

His adult life had been spent in naval service and scientific study of the ocean's currents. His efforts at charting the currents of the oceans was the most complete work on the topic to that date and is still considered one of the finest works of its kind. In 1859 he sent an inscribed copy of his *Sailing Directions* to Maximilian in Austria, which so impressed the young prince, who also served as commander of his country's navy, that he had the Austrian gold medal of arts and sciences conferred upon Maury. The friendship was further cemented the following year when Maximilian sent Maury the meteorological diary that had been kept aboard the Austrian ship *Elizabeth* on her voyage to South America.[15]

When the war came Maury resigned his commission in the United States Navy to offer his services to his native Virginia, and was later assigned to the Confederate Navy Department. His activities during the war consisted of designing and developing underwater torpedoes and mines, which the Confederate government never fully appreciated; and purchasing and procurement duties in England. He was in the West Indies on his way home from England when the news came of Lee's surrender. Believing that if he went home he would be arrested and imprisoned, Maury resurrected an idea he had had earlier in his career of establishing a colonization movement in Latin America. It seemed a perfect project for the refugee Confederates. "If Maximilian is wise," he wrote to his friend, the Reverend F.W. Tremlete, "and will encourage my plans I can assist mightily to make firm the foundations of his dynasty."[16]

Maury wrote to the Emperor from Cuba offering his services to the Empire, then left for Mexico without waiting for a reply. By the first of May, he had reached Vera Cruz and a few days later he was in the capital. Maximilian welcomed him warmly, listened to the colonization proposal, and offered him a cabinet post. Maury declined but did accept an assignment as Director of the Imperial Observatory.[17] Though favorably disposed to Maury's ideas of establishing an official office of colonization to assist Confederates who wished to immigrate to his country, Maximilian deliberated through the summer before giving an official reply. In a letter to Tremlette dated August 8, 1865, Maury told what prompted him to conceive the plan and what he expected to accomplish.

> In contemplating this shipwreck of country, kinsmen, and friends, I recognized among the debris

> of the wreck the very materials that are required to build, upon good and solid foundations, the Mexican Empire. Never, since the Revolution of the Edict of Nantes, has such a class of people been found willing to expatriate themselves. From such a wreck Mexico may gather and transfer to her own border the very intelligence, skill, and labor, which made the South what she was in her palmy days—except her bondage.

He later wrote that:

> A few of our clever farmers bringing with them their agricultural apprentices [former slaves], would give new life and energy to the country. By sprinkling the Empire with settlers of this sort, they and their improved implements of husbandry and methods of culture would serve as so many new centers of agricultural life, energy and improvement.[18]

Maximilian continued to express great friendliness toward the Confederates who were crowding into Mexico in spite of the fact that he had allowed the summer to go by without adopting Maury's colonization program. Finally in August the Emperor decided that such a program could perhaps help to bring stability to his country. He appointed Maury to the office of Imperial Commissioner of Colonization at a salary of $5,000 a year plus expenses, with the delegated task of assigning surveys, selecting lands for colonization and publicizing the movement in the United States. To assist in the survey of lands General John B. Magruder was appointed chief of the Office of Land Colonization, and former Missouri Governor Thomas C. Reynolds was named head of the American and Mexican Emigrant Company, chartered by Maximilian, which was to establish emigrant assistance offices in the United States and Mexico.[19]

On September 5, Maximilian issued his *Decree to Encourage Immigration.* (See Appendix B.) Embodying most of the fundamental suggestions made by Maury, the decree contained ten articles declaring Mexico open to "immigrants of all nations" and setting forth the general concessions, which would be made to all who came. Maximilian clearly intended to attract especially those Confederates who wished to escape the reconstruction of the South. Immigrants would be granted tax incentives, naturalization as soon as they had settled on the land, "liberty in the exercise of their respective forms of religious worship," and they would be free from any military service for a period of five years, though they would be allowed to form local militia organizations for their own defense.[20]

Maury drew up a set of "regulations and instructions" which were approved and published by Maximilian on September 27, 1865, and which were designed to elaborate on the original decree. The order classified the lands offered for colonization into three categories: (1) lands of the public domain that had never been under cultivation; (2) lands which had been under cultivation but which had for various reasons come under the jurisdiction of the government; and (3) private lands and haciendas offered for colonization by their private owners. Land distribution policies were also outlined and explained. Lands in the first class were to be donated by the government under an arrangement based on the United States land policy, in parcels of varying size up to six hundred and forty acres for heads of families.

The immigrants were to be divided into two classes: Class A included those who had lost all in the war, while Class B alluded to those who were not in "straitened" circumstances. The first class was to receive not only free passage by sea, but also a traveling allowance of ten cents per league to their new homes for each family member and "apprentices."[21] Class "A" immigrants would also receive land allotments in parcels of one hundred and sixty acres to a single man and three hundred and twenty acres to a man with a family, with a preemption right to as much more in each case. Immigrants in the other class could settle gratis on unimproved public land and were free to negotiate their own prices for private land. Furthermore all immigrants would be permitted to bring into the country duty free, all their personal effects as well as any implements necessary to their occupation or trade.[22]

Maury also prepared a lengthy "memorandum" to be published with the decree, which left little doubt that the entire colonization program was aimed primarily at one group: disaffected ex-Confederates.

> The Government not only invites all well-disposed persons to come and assist in the occupation of its vacant lands, but His Majesty the Emperor, touched by the spectacle of good men struggling with adversity in other lands, tenders hospitality and homes especially to these.
>
> Moved by the generous impulses of his nature, he offers them material assistance to enable them to reach this beautiful land. To those of them who wish to change their skies, make Mexico their home, and identify themselves with the country, a free passage by sea for their families and effects is offered.[23]

The colonization scheme also provided for the immigration of Southern Blacks. Maury felt, as did most of the Southern upper class that the former slaves had been betrayed by Northern abolitionists. Their situation in the post-war South was therefore taken into consideration in his program. "It would be difficult to say which have suffered, are suffering most," he wrote, "the whites from the ravages of war, or the blacks from the so-called kindness of their friends." He then outlined his proposals concerning their role in the immigration plan:

> Hoping to find for them sympathy in the heart of a generous minded Sovereign, and an asylum in his Empire, here am I, the advocate of Southern immigration, and a humane system of African emancipation in the United States, as the quickest, most certain and best means of affording relief to their suffering, of giving quiet to this country, stability to the throne, and peace to America.
>
> Though many of the Negroes have been set free...[they] are affiliated with both pestilence and famine, there are many of them such as are willing to accompany you to Mexico as apprentices, bound to serve as agricultural and other labourers...At the expiration of this term of service—say seven years—the apprentices will have earned a home as one of the rewards of his labour, and will be able to take care of himself.[24]

Maury reasoned that properly trained black laborers could do more for the economy of Mexico than they had for the South, being given the incentives of free men. These same laborers, numbering less than half the laboring class of Mexico, had been more than eight times more productive than their Mexican counterparts. Maury, therefore, had great prospects for the prosperity of Mexico. The introduction of "a few hundred thousand" of them could greatly raise the commercial output of Maximilian's Empire. Their value to the stability of the Empire would be inestimable, "guided as they should be, by the skill and experience of their former masters."[25]

The Northern press and some government officials, in their popular effort to strike out at the South, interpreted these statements as meaning that the Confederates would bring their emancipated slaves with them to Mexico and continue the old order without interruption.[26] Some of the New York papers accused Maury of plotting to re-open the African slave trade, an attack that he countered by saying,

> The Negro was set free in Mexico more than a generation ago. The Empire, the laws, and the people are all opposed to slavery; and anyone who could be so wicked as to desire to re-open the African slave trade might as well attempt to "plot" with the British Government, as with this, for that purpose. Nothing seems too absurd for the sensational press of New York.[27]
>
> Article 6 of Maximilian's *Decree to Encourage Immigration* stated, "Immigrants who may desire to bring laborers with them, or induce them to come in considerable numbers, of any race whatever, are authorized to do so, but those laborers will be subject to special protective regulations."[28] Further clarifying the Empire's position on the status of future black settlers, Maximilian issued a list of fifteen "regulations relating to black servitude which was published in conjunction with the Decree. The first article stated that all "person of color" were free by virtue of the fact that they were on Mexican soil. The document went on to describe the conditions under which such persons might be employed, or in Maury's terminology, "apprenticed." A black man could enter into contracts with an employer obligating himself to serve that employer for a period of "not less than five, nor more than ten years." The system as further described makes little distinction between slavery and peonage, which in many cases amounted to the same thing, except that at the end of his term of service he was free to go his own way with a certain percentage of his total wages held in trust for him in a local bank.[29]

Despite the enthusiasm with which the new Commissioner of Colonization applied himself to the immigration scheme, carefully executed arrangements and preparation had not been made. In one of his many treatises, which were circulated throughout the United States, North and South, Maury issued advice to prospective immigrants concerning the state of land availability. He wrote that the lands of Mexico had never been surveyed and that there had not until the present time been a land office.

> The consequence is that the government cannot tell which lands are public and which private, and though the chief of the land office is vigorously at work organizing surveying parties, and sending them forth into the field, it is recommended to those both in Europe and the United States, who desire to come to Mexico, to form themselves into companies, consisting of not less than twenty-five families each. Then while those at home are making preparations, let their pioneers come to Mexico for the purpose of purchasing a hacienda or other lands, of making ready to receive the rest.[30]

Maury himself was somewhat concerned but still optimistic. He wrote to his wife shortly after the Decree was issued, "The Decree has been injured in the translation. It is not as good for the State or the immigrations as I drew it. Still the papers proclaim it to be the most important act of the Empire."[31]

Notwithstanding, the colonization program went ahead and lands were surveyed near Cordoba, Guadalajara, Monterey, Durango, Mazatlan, Orizaba and Vera Cruz. The colonization fever seemed to affect all the Southern exiles. The *Mexican Times*, that began publication in mid-September, became the semi-official organ of the colonization office and printed all the immigration decrees and noteworthy colonization information. It also carried the Decree and Maury's memorandum on its front page. Henry Watkins Allen of Louisiana, the paper's first editor, outlined its editorial policy in the same issue.

> This journal will be devoted to the best interests of the Maximilian Empire. Its special

> object will be to advocate immigration and progress in their fullest meaning and extent…
>
> We shall urge with all our influence, immigrants from the United States and Europe, who wish rich, productive and cheap lands, to come to this country without delay, and accept the very liberal offers now made by the Imperial Government.[32]

Reflecting its Confederate bias, the paper issued a call:

> to all whose fortunes have been swept away by the terrible tempest that has so long raged in that afflicted land [the South],…to those who have drank the cup of bitterness to the very dregs we say come to Mexico. Here you will find liberty of conscience and freedom of Religious worship fully guaranteed. Here you will find a Monarch with expanded views and liberal ideas. In fact we may call him a Republican Emperor: for his daily labors are devoted to the good of his Empire, and the prosperity of his people.[33]

Within a few weeks of the Emperor's Immigration Decree the project appeared to be firmly established. Maury wrote to his wife on November 27:

> I am now selling those lands [near Cordoba] to immigrants as fast as they can be surveyed, at $1 the acre on 5 year's credit. There are about 40 of our people already there. Perkins has bought himself a house and sent for his family; so has Shelby, and so have a number of others. Mr. Holdham, an Episcopal clergyman, with his family-nice people-has been engaged by the settlement as pastor and teacher. I am going to reserve land for a church, cemetery, and school-house. Thus, you see, my sweet wife, colonization is a fact, not a chimera.

Excitedly he added, "By the time lands are paid for they will be worth even if no more settlers come to the Empire, $20, $30, or even $100 the acre."[34]

To further publicize immigration in the South, Maury was authorized to appoint seven colonization agents in the United States to publicize the movement and assist those desiring to emigrate. Agents were appointed in Virginia, Texas, North and South Carolina, Missouri, California, New Orleans and Mobile. These agents in a short time were deluged, as were the colonization office and The *Mexican Times*, with requests for information on immigration, available lands, planting season, and dozens of other topics.[35]

The question may understandably be asked as to why so many Southerners were anxious to leave their homes and all that was familiar to them to go into exile in a country with which they were, at best, vaguely acquainted. The answer lies in the state of mind of the Southerner, the condition of the South in the summer of 1865, and the attitude of some Northerners in high circles. The desolation wrought by the war itself had cast a shadow of hopelessness over the South. Men who had gone away to war four years earlier leaving flourishing acres of cotton and produce returned to ruined fields and dilapidated homes and farm buildings. Where there once had been a more than adequate labor supply there were now roving bands of newly freed Blacks aimlessly searching far and wide for the enigma of "emancipation," refusing to work and awaiting the fulfillment of the great "day of *Jubilo*." Besides the discouragement and dark mood brought about by the complete disruption of the life they had known, many feared that the wave of vengeance upon which Charles Sumner and Thaddeus Stevens were mounting would rise until it demanded the lives of at least the higher officials of the Confederate army and government.[36]

The fear of reprisal was no idle nightmare. Many former Confederate officials had already been arrested and were facing trial.[37] Those Confederates serving abroad when the war ended were excluded from pardon under the so-called Amnesty Proclamations, which were issued immediately after the war, and none of these officials could have returned to the United States "without the certainty of arrest, imprisonment, or, under the most favorable circumstances, the alternative of taking what has not been inaptly called the 'iron clad oath.'"[38]

Major Joseph Abney of South Carolina decided that he could no longer exist under such circumstances and determined to leave the country. Though his destination was Brazil, not Mexico, the feeling he expressed in a letter to a relative in Kentucky poignantly expresses the general pathos infecting the Confederates to whom reconstruction came hard.

> One can not (sic) close his eyes upon passing events, and situated as we are, entirely ignore the lessons of experience, and the solemn admonitions of history. The future is enveloped in clouds and darkness, and we were less than men if we made no efforts for the preservation of our families and to avert the manifold dangers that lie in our way…Our substance had already been consumed by the war…our people overwhelmed with debt, and sorrow, and disappointment…A deeper degree of destitution and want is inevitable, and as the negro will not work, and must eat, hunger, and starvation, and madness and crime will run riot through our borders and there is no earthly power that can interpose to save us and our children from the last extremities to which man can be reduced.
>
> I utter these sentiments with pain and sorrow, for I have long ceased to be angry, and there is no service I would withhold from my government, that might contribute to redeem and save our people. Laboring under these sad convictions, and being assured that vast numbers of our friends who are

> destitute of good farming lands, and now also destitute of labor, and yet are embarrassed with debt, must be brought to beggary, unless they exchange their accustomed habitations for a more genial clime and more fruitful soil, it is our purpose to form a Southern Colonization Society.[39]

The desire to leave unhappy conditions in the South did not wane as the yoke of reconstruction settled upon the old Confederacy. In February of 1866 The *Mexican Times* reprinted an article from the New York *Daily News*, which demonstrated that emigration was still very much a significant issue:

> The unsatisfactory state of affairs in the South, added to the great bad feelings exhibited by the Northern men and papers to Southerners, have made many of the latter conclude to leave the country. In several places whole families are assembling to emigrate to Mexico. Judge W. G. Swan, a member of the Late Confederate Congress from Georgia, is going to Mexico with 20 families. General John E. Williams, of Kentucky, left Nashville last week for Mexico with 60 Tennesseans. In Western Virginia 150 families are preparing rapidly to leave for Mexico.[40]

Several other newspapers recorded similar reports. The *Mexican Times* quoted articles from the *Lebanon Herald* and *Memphis Appeal*, both Tennessee papers; the Selma *Alabama Times*, *Charleston Courier* and several Virginia journals, all telling of groups either leaving or preparing to leave for Mexico.

Beyond doubt there were many among the expatriates who joined the movement for adventure's sake, and others such as Jubal A. Early who crossed into Mexico with a bitter hatred for the United States and swore that he would never return to the States "unless I can come back under the Confederate flag."[41] Regardless of their reasons for coming, they soon had cause to believe that they had closed a door behind them through which they could not return. The *Mexican Times* in November 1865 published the following in an editorial:

> We have positive information that all who have left the United States for Mexico are considered outlaws. In a letter from Mrs. General Shelby, of Missouri, to the General, now in Mexico, she says: 'General Sheridan told me that you had outlawed yourself by going to Mexico,'...We are also informed that every Confederate of note, 'who has returned to any Southern State from Mexico, has been arrested.'[42]

Happy to be free of an oppressive state of affairs in their homeland, and certain that the United States would never again extend to them their former rights and privileges as citizens, the Confederate expatriates determined to reestablish the old life in their new, adopted country.

CHAPTER SIX

The Colonies

Matthew Maury quite accurately described the state of affairs in the fall of 1865 when he stated, "colonization is an established fact." It seemed to North American visitors in Mexico City that ex-Confederates were everywhere. The Colonization was besieged with requests for information concerning immigration and available lands. Surveying parties were being recruited to examine and map the land set aside for colonization. *Haciendados*, private Mexican landowners, were offering portions of their holdings for sale throughout the Empire on very liberal terms. In short, the prospects for the Confederate colonization of Mexico seemed brilliant indeed.

The overzealous Maury and others continued to disseminate accounts, which would press Sir Walter Scott in the superlatives they used to describe the climate, the land, and the "most favorable conditions." They saw in Mexico "a land more blessed by nature in its soil and climates than any part of the United States;" a land where Adam and Eve tended the Garden and David sang psalms to his thriving flocks. Many examples from Maury's letters and notices could be cited that would perhaps explain the vision with which the immigrants arrived in Mexico, but the following will suffice:

> The earth here yields to the care of husbandry with a profusion that would seem incredible there [in the United States] and fabulous Europe. In some places it crowns the labor of the husbandman regularly with two, and in others three, harvests annually; and in each one he gathers one hundred, two hundred, sometimes three hundred, and occasionally four hundred fold, and even more, according to his own skill and the kind of seed used…
>
> Under these fine climates, which give a purity and transparency to the atmosphere that make existence itself an enjoyment, and invest the eye with the faculties almost of a new sense, the vegetable kingdom displays its wealth and its powers most gorgeously, and with the most marvelous vigor and concentration.
>
> In chosen spots, and upon a single hacienda, may be seen crowded together, piled up in steppes one above another, in all the glory of the lily, and with the perfection of excellence--fruits, flowers, and productions that, in less favored climes, require for their display as many latitudes, climates, and soils as can be found in the entire breadth of plain that lies between the sources of the Mississippi and the mounts of the Amazon…
>
> The Mississippi valley, even in its palmiest days, could not boast any plantation that could compare in baronial splendor, lordly magnificence,

> and princely hospitality with your Mexican hacienda...
>
> He [the immigrant] may pitch his tent on the verge of highly cultivated districts, from which he can draw his supplies until the bountiful earth, yielding to his own good husbandry, shall yield him of her increase. And this the soil of Mexico, under climates that have no winter, will do in two or three months.[1]

Thus did the gospel of Mexican Colonization stream in to the South, making hundreds, perhaps thousands of converts.

By November 1865, interest had become so great in the colonization project that the American and Mexican Emigrant Company was organized and chartered by Maximilian and authorized to secure land for colonization and assist immigrants in obtaining transportation to Mexico and in locating land for settlement. By far the most important of all the immigration companies organized during this period, the American Mexican Emigrant Company appointed agents in principal cities in the United States to disseminate information about Mexican land, climate, resources, and the privileges accorded by the Empire. Under its charter the company was exempted from the five percent land transfer tax on all land secured for colonization purposes and allotted to settlers. Parcels of land sold to immigrants could not exceed one square kilometer and must be offered at a "fair price."[2]

The American and Mexican Emigrant Company was not the only one authorized by the Emperor. Maximilian had declared Mexico "open to immigration from all nations" and several efforts were made to bring settlers from Asia as well as from Europe. Dr. Manuel B. de Cunha Reis was appointed to head the Asiatic Colonization Company in the fall of 1865, with authority to bring Chinese laborers to Mexico to assist with railroad and public works construction. Reis had other interests, which occupied his attention leaving less of his time to attend to the affairs of his colonization business. The company, therefore, did little, if anything, to fulfill its contract. Maximilian revoked its charter early the following year, but Reis obtained another appointment, contracting to bring 8,000 Portuguese settlers to occupy government lands at Mitlaltoyca and Amixtlan in the Huachinango district. He approached this assignment with the same enthusiasm he had for Oriental immigration. Eventually a few Americans became interested in the lands intended for the Portuguese, but only a few of them settled there.

European immigrants were, next to Southerners, the most desirable class of prospective settlers during this period. By Imperial decree Numa Dousdebes, a Mexico City merchant, was commissioned on October 18, 1865, to organize a company in Paris to bring French and Spanish colonists from the Basque Provinces. Though agents were appointed to recruit immigrants, few Frenchmen, seeing the shaky hold the Franco-Mexican party had on Mexico, were willing to leave their secure homes for a new land, and so the program ended in failure.[3]

The initial success of the American and Mexican Emigrant Company is perhaps attributed to the fact that conditions in the southern United States had caused disaffection and discouragement among so many former Confederates. This company, headed as it was by former Confederates, also had the unqualified support of General Magruder's Office of Colonization and, no doubt, had advance notice of available lands and first choice on the most desirable of that. Not only was the public land surveyed under the direction of that office, but also all private land offered for sale was first registered with this agency.

The first area pronounced available for settlement was in the Cordoba-Orizaba-Vera Cruz territory. This is the region where most of the Confederate colonies would be laid out. The country around Cordoba is surrounded by mountains with an average elevation of three thousand feet above the Gulf, with a climate ranging from seventy to ninety degrees Fahrenheit. On the east are the Chicahaita Mountains and on the west lie the Tierra Calientes. To reach the Cordoba Valley one must travel westward on the road from Vera Cruz, and then through Paso del Macho, a narrow gorge that had been created by the Rio Atoyac. The walls of the canyon were described as "almost perpendicular on both sides and more than a thousand feet deep. A fine wagon road was constructed through the pass under the authority of the Spanish Viceroys [during the eighteenth century] by cutting into the side of the mountain and walling up on the lower side, the wall in many places being over one hundred feet high."[4]

It was said that Hernan Cortez when on his way to Mexico City named Paso del Macho, or Mule Pass, during his campaign against the Aztecs in 1519. It was also rumored that as Cortez' army camped near the pass, a pack mule came up missing and the Spaniards sent to find the mule stumbled upon it on the other side of the mountain. There was revealed to them a panoramic view of the canyon and the beautiful valley beyond.

The Spanish were the first to colonize the area. They were as taken with the majestic beauty of the Cordoba Valley as were the later Confederate colonists who often referred to it as a paradise rivaling the Garden of Eden. However, during the Mexican Revolution with Spain, destruction swept all the first colonists away leaving scores of once grand haciendas now reduced to moldering ruins and overrun by jungle growth.

Long before the Confederates began arriving there in 1865, clusters of ruins dotted the valley, testimony to man's temporary footprint upon the land. And the Confederates would learn that same lesson of impermanence through bitter disappointment as their time in the spectacular Cordoba Valley quickly passed. To them the realization would come too soon that the Valley was not the Garden of Eden but rather a *Paradise Lost.* Their memories and crushed dreams still haunted the region, along with those of that earlier wave of colonizers who had been lured by its intoxicating beauty three centuries before.

However, in that first flush of visionary excitement, the Confederates looked to the future, not the past. The site for the first village was selected at a point about seventy miles west of Vera Cruz, and nine miles southeast of Cordoba, near the present village of Paraje Nuevo.[5] The town and surrounding plantation tracts were marked out on 500,000 acres taken from the Catholic Church by Juarez and later "ex-appropriated" by Maximilian and applied to colonization. Maury named the colony *Carlota* in honor of the Empress who had taken much interest in the material welfare of the exiles.[6]

The new town of Carlota was only three miles off the Imperial Highway that ran from Vera Cruz to Mexico City, but the Imperial Highway was completed only as far as Paso del Macho, about ten miles from the outlying colonies. By the spring of 1866 in addition to Carlota there were thriving colonies at Orizaba, Cordoba, Omealco and eventually, in November 1866, at Tuxpan, near the coast some one hundred and forty miles north of Vera Cruz.

Judging from the reports of the first settlers in the Cordoba region, even Maury's exaggerations were credible. A letter to Governor Allen from "a distinguished gentleman" who had been examining the lands near Cordoba expressed great satisfaction with the area and visionary expectations for settlement there. He wrote, "There is no portion of Missouri, Kentucky, Louisiana, or Tennessee that surpasses it. Coffee, corn, cotton, tobacco, and every variety of fruit grow as well as they are produced in any country in the world."[7] William Marshall Anderson, a younger brother of the Major Robert Anderson who had commanded Fort Sumter when the opening shots of the war were fired in 1861, had come from Kentucky to Mexico in May to serve Maximilian as a surveyor. Shortly after his arrival he passed through Cordoba and was impressed by the profusion of vegetables and fruits available in the market:

> But what is there in the market? One might think it would be easy to tell in so small a place as Cordoba, but it [is] far easier to ascertain and

> designate what there is not. The fruits and vegetables, the quality, quantity, and variety is truly amazing. There would seem to be specimens of almost everything[,] which ripens in or grows to perfection above the ground.[8]

General Shelby, who along with General Price and several other interested Confederates, took a three-day tour of the Cordoba area and wrote his impressions to a friend:

> Let me say to you, I was raised in the best part of Kentucky [Cordoba] I have ever seen...I only regret that I am without language to describe it as it should be...a fortune awaits any man who owns a farm in this Valley, and who will exercise economy and energy for a few years. Don't think I am extravagant—You will agree with me . . . [9]

It is little wonder then that the colony began to flourish from the beginning, and lying as it did on the line of the uncompleted Imperial Railroad from Vera Cruz to Mexico City and beside the principle road to the capital, Carlota was "the gateway of colonization." General Sterling Price came to be looked upon as the leader of the colony, and within a year owned a prosperous coffee and cotton plantation. Besides General Price, Carlota claimed the residence of some of the most illustrious Confederate officers of the late war. General James E. Slaughter and Captain Henry Price, the General's son, went into the rich timber belt between Cordoba and Orizaba and put into operation a steam sawmill in addition to growing cotton.

Several professional men such as Doctors Xaupe and Russell, and General Walker H. Stevens, one of General Lee's chief engineers, engaged in their former occupations. The dauntless Shelby became a farmer and entrepreneur in partnership with Major McMurty, a prosperous freight contractor who ran a wagon train and stage service from the railroad terminus at Paso del Macho to Orizaba. The partners also secured a concession to transport iron and other railway materiel to Mexico City. Former governors Isham Harris of Tennessee, and Perkins of Louisiana were also there, and with Price formed the vanguard of the plantation builders who settled at the Carlota site. Generals Hindman, Hardeman and Watson also established smaller agricultural projects along with Colonels Fox, Green, Roberts, Whaley, and others of lesser renown.[10]

As the winter of 1865 gave way to the spring of 1866, a sense of security and prosperity settled over the booming colony project. A reporter from the *New York Herald* traveled to the colonies during this time and wrote the following:

> The village [Carlota] was firmly established with a population of thirty families, laid out systematically around a large square, which is covered with a natural growth of trees, among which is a grove of mangoes, the most beautiful shade trees, whose interlacing boughs and thick, glossy foliage exclude completely the sun's rays, affording a delightful retreat during the heat of the day. In a few years, Carlota, will be the most pleasant town in Mexico. It will be built after the American style, with a degree of beauty and comfort elsewhere unknown in the Empire and the society there will be formed from the best educated families of the South and West . . .[the ubiquitous reporter, who remains unidentified, could not have made a more inaccurate prediction on the future of the colonies, however, he went on to say] Wages are low, compared to what are paid in the States, laborers receiving seventy-five cents and one dollar a day and board; but with this a man can soon secure a fortune...Forty acres in coffee is a fortune to any man, and there are a dozen other articles that can be raised with little labor, and which will pay 200 per cent more than the best crops that can be raised in the States.[11]

With the success of the Carlota colony so apparent, Maury, Governor Allen, and other promoters of Confederate settlement, increased their colonization efforts. Convinced that Confederate colonization would fulfill the earlier French dream of regenerating the country and advancing it along the path of progress and stability, Allen editorialized in November, 1865:

Immigration will, in a great measure, destroy the most ruinous system that has ever cursed the agricultural interests of any country—we mean absenteeism...

Finally, we advocate colonization for this and a very important reason: There are in round numbers 8,000,000 of human beings in the Mexican Empire--of these 1,000,000 belong to the upper class and 7,000,000 to the lower class. There is no middle class in Mexico. All political writers- all statesmen of every land tell us that the middle class is the bulwark of the State...Immigration will make a middle class. It will bridge over the chasm now existing between the upper and lower classes...This journal will use all its influence to ameliorate the condition of the toiling millions in Mexico.[12]

Maury shared these sentiments and claimed that under such an enlightened monarch as Maximilian the colonization movement would make Mexico a beacon to the civilized world. Maximilian had already undertaken the task of educating and improving the condition of the laboring class to which colonization was to be such a blessing. Maury wrote:

Many good men of the country look upon immigration on a large scale as the readiest and best means of restoring the equilibrium of the classes, and of giving to the country and its institutions that stability and force which are so essential to the full development of its vast powers, capabilities and resources.[13]

It is impossible to determine how many were reached by the colonization information disseminated by Maury through his immigration agents in the United States, and through The *Mexican Times,* but judging from the widely separated newspaper accounts of groups and individuals preparing to immigrate to Mexico, the number must have been sizeable. For every immigrant who actually arrived in Mexico there must have been dozens more who had read the colonization literature and given emigration serious thought. No doubt such deliberation and careful planning were largely results of ominous and timely warnings from friends and local newspaper editors. Few Southern papers encouraged their neighbors to emigrate, though most editors freely admitted that conditions in the South justified a resort to "extreme measures."[14]

Wade Hampton of South Carolina publicly discouraged emigration and called for all Southerners to stand together in trying to better post-war conditions in their native land. A letter from Robert E. Lee to Matthew Maury represents the feeling of many of the Southern leaders who chose to remain in the South despite the unhappy state of affairs.

> We have certainly not found our form of government all that was anticipated by its original founders; but this may be partly our fault in expecting too much, and partly due to the absence of virtue in the people. As long as virtue was dominant in the Republic, so long was the happiness of the people secure.
>
> The thought of abandoning the country, and all that must be left in it, is abhorrent to my feelings, and I prefer to struggle for its restoration, and share its fate, rather than to give up all as lost...I shall be very sorry if your presence will be lost to Virginia. She now has need of all her sons, and can ill afford to lose you.[15]

No one can say how many prospective emigrants were influenced to remain in the South and "share its fate." Nor is it possible to accurately determine the number of settlers who came to Mexico seeking new homes and a fresh start during the most energetic period of colonization, which lasted from the summer of 1865 until the early fall of 1866. We may count names and estimate the number of family members, but it would be fruitless, for some became discouraged and homesick the moment they arrived and returned home almost immediately. Others came to prepare the way for their families, but were never able to bring them to their new homes. The total number of colonists varied with the seasons, with the months, even with the days. One settler estimated that at one time about 500 colonists made up the Carlota colony, but even this is open to question.[16] A list of arrivals at the port of Vera Cruz appeared periodically in the *Mexican Times* from which the following chart is compiled.[17] It should be remembered that the arriving ships represented were not given over completely to the transportation of prospective settlers, but were regular commercial or passenger vessels engaged in routine commerce. Therefore not all of their passengers were colonists. Furthermore colonists came overland in great numbers[18] or entered Mexico by sea through Tampico to the north and Guaymas located on Mexico's west coast.

NUMBER OF FOREIGNERS ARRIVING AT VERA CRUZ SEPTEMBER 1864 THROUGH DECEMBER 1866

DATE	YEAR	TOTAL	AMERICAN
September	1865	264	29
October	1865	290	---
January	1866	239	---
February	1866	765	540
July	1866	186	11
October	1866	252	30

December	1866	81	---
	Total	2,077	610

Although the primary emigration effort was made in the South, all Americans were invited by Maury's agencies to come and share in the "noble task" of bringing Mexico into the modern age. When one gets past the list of original colonists, it is difficult to identify as a Confederate every settler who is referred to as an American. It is safe, however, to assume that most were Southerners since they departed in larger numbers from Southern ports or entered overland through Texas, and once in Mexico, almost unanimously chose former Confederate officers as their leaders and representatives. That there were also former Union soldiers in the predominantly Southern settlements is borne out in an article that appeared in the *Mexican Times* under the headline: "Some Confederates Badly Reconstructed." After discussing the general observation that, "with few exceptions," the Confederates in Mexico still considered themselves American citizens, exhibiting less feeling over the failure of their cause and loss of property than might be expected, the article went on to say:

> Generally they treat Northerners with the utmost kindness and cordiality, although between them the causes and occurrences of the war are a tabooed subject. The most considerate and sensible among them are from South Carolina and Missouri; while among Tennesseans and Texans still lingers a little of the old leaven which too often shows itself in the inter-course of the younger representatives of those States with Northern men.[19]

New arrivals joined with the older settlers who had made Carlota a thriving community in proclaiming the colony "the garden spot of the world." General Price was greatly excited about the agricultural prospects and in a letter to the *Missouri Republican* he issued an encouraging call for further settlement.

> When the character of our lands is well understood, immigration will be a fixed fact under any circumstances, and the finest lands, that can now be procured at low rates, will command large prices. I have never known the cultivation of lands to yield such large profits. My neighbor, Mr. Fink, (a man of science,) cultivates eighty(80) acres in coffee with ten hands, and sold his last year's crop for $16,000. His coffee-field, shaded with every variety of fruit trees, in full bearing, and the walks fringed with the pineapple, is the most beautiful sight I have ever seen."[20]

But other less populated areas had in the meantime been attracting a great deal of attention. Even as the first settlers in the Cordoba region were pitching their tents, other lands were being surveyed and readied for immigrants. From the Rio Grande southward to the State of Oaxaca, and from Tampico on the Gulf of Mexico to Guaymas on the Gulf of Cortez, representatives of the Office of Land Colonization and private speculators were selecting choice parcels for future immigrants. A hundred and twenty miles north of the Carlota settlement lay the Valley of Tuxpan, an area ranking with Cordoba in its beauty and fertility.

Shelby, though possessing land in Carlota, was also interested in the prospects for colonization in Tuxpan. With several associates he established an organization to publicize the area and to assist in bringing settlers to the region. Several months after the founding of the first Confederate village at Carlota, the new colony was off to a "splendid start" and prospects appeared "cheery" for

the Tuxpan venture. Land shares had been on the market for some time with many already taken. Shelby's company intended to buy "schooners" to be used in transporting to Tuxpan free of charge, all settlers who were unable to pay their own passage.[21]

Tuxpan promised to outshine the Cordoba region in its suitability for colonization and became equally as important as Carlota during the peak of the movement. But these were not the only lands open to settlement, and the growing number of Confederates seeking homes in Mexico required the opening of new areas. By the spring of 1866 there were small communities of Southerners in or near the cities of Monterey, Saltillo (which had once been the capital of Texas), and near San Luis Potosi. There were also the beginnings of small colonies in several other districts that had been newly opened.[22]

Available public lands began to disappear early in 1866 as the first rush of settlers eagerly took up the surveyed and posted land. The Office of Land Colonization, burdened by the Emperor's lack of attention to the project due to a depleted treasury and increased pressure from the Juaristas, was desperately trying to get completed survey reports approved by the Imperial Minister de Fomento (of Improvement) and to organize new survey expeditions. Fortunately many private Mexican landowners had offered to sell or lease portions of their holding to Confederate settlers and these lands, eagerly bought and rented by speculators and former Southern planters hoping to revive the old way of life, helped to absorb many of the excess settlers. Mexican landowners whose haciendas lay in or near those areas controlled by the Liberals were anxious to attract the Southerners to their estates to serve as part-time militia in defending their lands against the incursions of Juaristas, bandits, and Indians.

One such offer was made by Buenaventura Saravia, proprietor of the vast hacienda of Chaparrion, in the neighborhood of Nombre de Dios and Parilla. To each colonist and his family Saravia promised, "gratis for five years," a *fanega* (1.6 acres) of fertile land, a yoke of oxen, two cows with their calves, two mares, one horse, two ewes, two hogs, two goats, one jackass, one chicken cock and two chickens, and in addition to all this, "the necessary tools for planting." Furthermore he would give two measures of corn, one half measure of beans and one dollar per week.

> If at the end of the five years, the colonists have realized fine profits, they will pay me the price of

> the animals and the tools. We will then make a new contract for the future, in which, for my part, I will bring perfect good faith. I only ask for the present of my colonists honest, good conduct and the obligation to assist me in defending my property in case of attack by Indians or robbers.[23]

A Reverend Mitchell, a Confederate colonel from Missouri, who had come to Mexico among the first arrivals, arranged a ten-year lease on one of these haciendas. The estate amounting to 500,000 acres was located in the beautiful valley of the Rio Verde, midway between San Luis Potosi and Tampico. Mitchell invited fellow Confederates to the colony, but with moderate success.

Other lands were opened up all over the Empire, which enjoyed more and less success than Mitchell's enterprise. In the State of Chihuahua a colony was established by a Mr. Bryant from Arkansas to which settlers were invited, first to rent the land upon which they settled with the privilege of purchase at the stated price. General Terry, of Texas, had begun a small venture in Jalisco and later went into partnership with General Hardeman on the west coast in the Guadalajara section where they purchased 22,000 acres to be opened up for settlement. In Tamaulipas, on the Panuco River thirty miles north of Tampico, D. J. O. Forns owned the entire Hacienda de Limon amounting to twenty-five square leagues, but was able to attract but very few settlers since his estate was too close to Liberal-held territory. In addition to these, A. J. Grayson had secured land in Mazatlan and petitioned the Emperor for title. In addition at least one settler, Judge D. O. Shattuck, was growing his first cotton crop thirty miles from the town of Mazatlan in 1866.[24]

By the fall of 1865 there were nearly ninety-eight families living on homesteads and by the first part of 1866 that number would jump to nearly five hundred families. The colonies at Carlota and Cordoba were more promising than the other smaller towns such as Orizaba, where James E. Slaughter and Captain Herbert Price (General Sterling Price's son) started a steam sawmill along with General Hamilton Bee who planted a cotton crop. At Orizaba as well as Cordoba and Omealco, the native competition for land and crops was severe. Local Mexican merchants did not welcome those Southerners who entered into commercial enterprises. Nevertheless, the Confederates worked the soil, planting cotton, sugarcane and coffee, and attempted to cling to their old traditions.

They organized hayrides, picnics and square dances. There were even some literary circles for ladies who, in some cases, imported their own pianos and started group-singing clubs. Most of these folks envisioned a transplanted Dixie homeland with a growing economy that would eventually have schools, colleges and their own newspapers. It was all wishful thinking by these dispossessed relics of a lost culture.

It was only natural that in such a wide-open immigration program, someone would try to take unfair advantage. For example, the Palacio Colony in the State of Durango had been planned as a model community for Mexico and Edward Conway, of Missouri, was sent to survey and lay out a suitable section for the village. Having done this, he somehow got title to the best lands, withheld them from settlement, and planned to speculate at enormous profits. In April 1866 George W. Clarke was appointed agent of colonization for Durango and immediately upon his arrival realized what Conway was doing and exposed him. By the time the survey was resubmitted to the Imperial Office of Improvement and the title cleared, Juarista activity throughout Mexico had so occupied the attention of the Empire that only a few Americans ever settled there.[25]

By the spring of 1866, Maury had appointed land agents for all the areas designated for colonization. Their primary duties were to assist new arrivals in locating suitable land and, as in the case of men like Conway, to protect the settlers from misrepresentation. Colonization agents for some of the most promising districts were Y. P. Oropesa at Vera Cruz; J. Perkins at Cordoba; J. T. Lux,

Monterey; Alonzo Ridley at Mazatlan, and D. Ramon de la Vega at Manzanilla.[26]

All the colonization plans had as their primary objective the relocation of Southern planters and farmers who had either been ruined by the war or felt it unbearable living under an oppressive reconstruction program manipulated by the vengeful radicals in Congress. However every such settlement also required business and professional men and other services such as ministers, teachers, and blacksmiths. In the South was an ample supply of these who shared the feelings of the agriculturists, and they, too, came in great numbers to the Mexican colonies. But since at the beginning the need for these occupations was not as great, such opportunities were soon filled and professional men, with no knowledge of the soil, found themselves starving in what promised to be a harvest of prosperity. A warning was published in one of Governor Allen's editorials that the colonies were "no place for doctors, and lawyers, and clergymen. These vocations are filled to overflowing . . . business here is also overdone, and we advise no adventures of this kind."[27]

The Imperial Commissioner of Colonization, recognizing the extent of this problem began officially to discourage "for a time" the emigration of professional men who "have no knowledge of agriculture." This policy did not mean, however, that colonization was proving itself a "humbug" and that settlers looking for good farming land should not come to Mexico. The familiar propaganda still went out to North America. "Wealth is accumulated here faster than perhaps in any other portion of the world," the literature boasted. Prospective immigrants could begin on a small amount and "find at their door" a sale for all they could raise "without any expense whatever" in marketing their produce.

In all fairness to Governor Allen, on the other hand, it should be said that he made an attempt to be realistic in his appeals for colonists. "Do not come without capital," he wrote. "Emigrants aiming to Mexico must not expect to sleep on a bed of roses. They must come prepared to submit to an immigrant's life."[28] But come they did, by ship and wagon and on horseback, with little thought to the fact that they were trying to escape the memory and aftermath of one civil war by fleeing into the midst of another.

CHAPTER SEVEN

Decline of the Colonies

With the close of the American Civil War the days of Maximilian's Empire were numbered. United States Secretary of State Seward had been under pressure throughout the greater part of the war to take action against the French intervention. But preoccupied as he was with the problems of war in his own country, Seward had been able to issue only an occasional protest note. With the North triumphant over the Confederacy it now became possible for Seward to take a more decisive attitude toward the French adventure. General Sheridan in seeking permission to march into Mexico was only echoing the popular demand. Seward, however, believed that assisting Juarez in the form of arms and money could gain the best effect at the smallest expense, at the same time leaving the French to speculate as to American intentions.

Juarez knew, even if the portent was not as clear to Maximilian and Napoleon, that the Federal victory in the United States would have great significance for his cause. Though he felt confident that his troops, without any foreign assistance, would "wear out the French and compel them to abandon their iniquitous enterprise," he was pleased that Lincoln had refused to recognize Maximilian's regime. "Napoleon," he wrote from his provisional headquarters on the Rio Grande, "may be meditating a new twist to his interventionist policy," but at any rate Juarez was certain that the attitude of the North would "spread great discouragement" among the French. "Hence I judge," he continued, "that they are already reaching the period of their decline and that the reaction of the peoples against their oppressors is beginning."[1]

In his attempts to please all parties Maximilian vacillated dangerously between decision and reconsideration as he gave heed first to one advisor then another. In August 1865, Maximilian mounted an expedition to the cities of Queretaro and Guanajuato, during which the staged demonstrations of the populace and feigned enthusiasm of his advisors convinced him that he was truly popular with the people. While on the same expedition he was erroneously informed that Juarez had abandoned his cause and Mexico.

Tragically he decided that the Liberals were defeated and that any remaining resistance was the work of mere dissidents to whom decisive defeat meant nothing, but who must be exterminated before peace could be secured. He therefore issued on October 3, 1865, the infamous "black decree" ordering that any armed prisoners taken in the act of resistance to Imperial troops were to be shot on the spot.[2] When executions were carried out under authority of this decree, many heretofore uncommitted Mexicans rallied to Juarez, and the Liberal movement became more powerful than ever.

It is curious that the Confederates who had proven themselves among the most capable and outstanding military strategists in the world were not able to accurately interpret the ominous events occurring in Mexico in the fall of 1865 and early months of the following year. Few of them upon their arrival there in the summer of 1865 had any illusions concerning Maximilian's military state of affairs. In fact it was the Emperor's dependence upon French and poorly trained native Mexican troops that had caused men like Shelby and Kirby Smith to believe that the Empire would accept the services of Confederate soldiers. The Convention of Miramar[3] had provided for the eventual withdrawal of French troops, and it was suggested by some Confederate generals that a large American force be recruited which would serve as an elite guard to augment the Mexican Imperial Army, consolidate the government against the time of the withdrawal of the French soldiers, and hold the country until the people became reconciled to the change.[4] The Emperor for obvious reasons declined the Confederate offer, but the American press, receiving most of their information from United States government sources, continued throughout the peak of colonization to accuse Maximilian of using such troops in his army.[5]

Instead of foreseeing the failure of the Empire, and consequently their own plans for colonization, the Confederates remained on friendly terms with Maximilian's government, some of them even publicly supporting it in assailing Juarez. If there was a consensus among the Confederates toward Maximilian it was probably best expressed by William Marshall Anderson:

> I have no doubt that the whole world would be benefited by the establishment of the present just and generous government—the people of the United States in an especial manner. I have...little doubt that if the republic is revived, with it will also be revived disorder, confusion, and anarchy.

Though, being Americans, they were traditionally more favorably disposed toward a republican form of government, they seem to have been convinced that Mexico was incapable of such government and desperately needed a monarchy that could bring peace and prosperity instead of endless revolution and ruin. To them the Juarez government epitomized the anarchy and bloodshed that had afflicted Mexico throughout the nineteenth century. In one of Governor Allen's editorials he did little to win the friendship of the Liberals, which would later be sorely needed in the event the Empire should collapse. Without French troops to maintain order and protect the Empire the Southerners would be left alone in Mexico. Further alienating the Americans from Juarez and his supporters one Allen editorial told its readers that Juarez "is placed in Chihuahua, issuing decrees and *playing would be president* (emphasis added)."

> He was induced to believe that the Imperial troops had entirely abandoned the frontier departments of the Empire, and he, poor soul, came all the way from the quiet village of El Paso [now Ciudad Juarez] to receive the 'acclamations' of the good people of Chihuahua, whom he had so recently robbed by his forced loans…
>
> What a farce! This self-styled President, whose term of service has long since expired, who has no more claim to the Chief Magistracy of Mexico than any other of the eight million of her inhabitants, still keeps up his mock court and plays the saddlebags dictator. He has no party in Mexico. His only sympathy comes from the filibusters among our neighbors of the North.[7]

In addition to their attitude toward the Liberal government, many of the colonists had a low opinion of the Mexican citizen, as well. Unfortunately most of the Confederates, whose statements concerning the Mexican character are on record, looked upon their hosts as lazy, ignorant, and unwilling to improve their conditions.

Jubal Early, it was said, only spoke two words of Spanish, "and those contemptuously, because the words themselves expressed so aptly the Mexican's idea of eternal procrastination."[8] Judge Terry admitted not long after his arrival that he was "not in love" with the Mexican people. His reason was that "like most of those in countries where the earth produces abundantly" they were indolent and ignorant.[9]

Perhaps these exiles should not be judged too harshly in the light of the general Mexican attitude toward them. They came to Mexico at the invitation of an unpopular government, settled on lands confiscated, in many cases, from those who were with the Juaristas, and considered themselves citizens of a country whose language they could not even speak. Besides they were *gringos*, a word that in its very Spanish pronunciation evokes contempt; representatives of a nation that had taken all of Mexico's territory north of the Rio Grande only a few years before. Mexicans never had much faith in *Norte Americanos*. They looked upon them as grasping and ambitious, covetous and unprincipled. William Marshall Anderson, perhaps because of his Catholic faith, gave more attention to the Mexican people than most of the Confederates, and understood the jealousy and fear with which the exiles were regarded. "They were oppressed by the Old Spaniards, robbed by the Americans, and they are now being run over by the French," he wrote.[10]

Although the Confederate colonists had sympathies toward the Empire, they claimed to be neutral in the current struggle. According to the *Mexican Times* the Emperor had welcomed the Southerners but requested that they "take no part in the politics or internal feuds of the empire, but at once settle down in quiet as became good citizens."[11] Except for the offer of some former Confederate soldiers to enter Maximilian's military service, the records show that they did just that, though they were continually answering the charge of fomenting trouble with the United States. "We are glad to see and to say," Governor Allen wrote in answer to the charges, "that among the Confederates who are now in

Mexico, there are no secret juntas conspiring against the land of their birth--no loud denunciations against [the] Fatherland."[12]

The Fatherland, however, was concerned with only one thing at the time and that was to get the French out of Mexico and the Constitutional Republic back into power. Since the Confederates were colonizing in Mexico with the blessing of the French-backed regime and without permission from what Washington considered the legal government, they were included in that aim. Sympathy for former Confederates, especially those who had taken residence in a foreign country, was singularly absent in Washington's halls of government. Increased speculation concerning intervention by the United States brought up the question of the Monroe Doctrine once again. The Liberals, for the time being, were happy to accept the American interpretation of the doctrine, while the Imperialists felt that the United States was guilty of the very thing which they were accused of: interfering in the affairs of Latin America. Shortly after his appointment to survey lands for colonization, Anderson observed:

> The loud outcry in the U.S. against France for infringing upon the Monroe Doctrine is nothing but national Pharaseeism, an arrant hypocrisy. By professions of friendship and sympathy, she wishes to encourage Mexico in her ridiculous efforts to establish a republican government, knowing full well that will result in her biennial revolutions and pronunciamentos, and that thereby and thereupon she, the elder sister, will take another broad slip from the Mexican gown. She has robbed Mexico, by force and violence, robbed her of Texas and California [sic], and now her mouth is watering for a straight arc from below Metamoras [sic] to the lower point of the Gulf of California...[13]

The *Mexican Times* also joined the controversy with an editorial accusing the Untied States of opportunism:

> Why then this new born zeal for the Monroe Doctrine which has been long since exploded, and which no two men understand alike in the United States? It is now…the guise of patriotism as a political claptrap for party purposes…
>
> Many a cunning politician has an axe to grind. Many a loafing, roving, filibuster would like to enlist for plunder. The Juarez bondholders and their sympathizers--these are all 'Monroe Doctrine' men, and swelling to burst with unselfish unadulterated patriotism, and who can see no good in any government on earth except a republic…
>
> The Monroe Doctrine is a bloated humbug. Its gas bag will be punctured by the good sense of the American people...[14]

On February 12, 1866, Secretary Seward demanded the withdrawal of the French forces, and Napoleon was faced with the final realization that his adventure had no chance of success. Perhaps hoping that Seward was bluffing and would not use United States troops to force his withdrawal Napoleon delayed, then gave a temporary promise to withdraw his troops.

Juarez, on the other hand, was greatly encouraged by the more militant tone of Seward's note, and began to speculate on United States intervention. Just before the end of the American Civil War when a Northern victory was certain he had written:

> if that Republic [the United States]…as a friend and not a master, should wish to lend us some assistance in soldiers or money, without exacting humiliating conditions or the sacrifice of a single inch of our territory or any decrease of our national

> dignity, we would accept it, and secret instructions to that effect have been sent to our Minister [Romero].

Second thoughts later occurred about the actual use of American soldiers. Since Americans were not as accustomed to poverty and hardship as his soldiers, such an army would have to be extremely well paid and supplied…

> otherwise it would become a plague by its subordination and errors, and in that case the cure would be worse than the disease. Hence those persons who have applied for permission to bring volunteers…have been met with the condition that they provide their own resources.[15]

Juarez would not have to concern himself with the problems of American troops in Mexico. Napoleon, shaken by Seward's demand and the presence of Sheridan's troops along the border, instructed Marshal Bazaine, as commander of the French troops in Mexico, to mount a vigorous effort to crush Juarez after which he was to prepare to evacuate all French troops.[16] When Bazaine's efforts proved insufficient he began to retreat, withdrawing his armies from Monterey and Saltillo and Tampico. The Liberals, with increasing strength, followed.

The hostility of the Mexicans toward foreigners increased in direct proportion with Liberal gains throughout 1866, putting even more pressure on the Office of Land Colonization in its effort to successfully settle the immigrants who continued to arrive in Mexico. The fantastic propaganda that Maury had been pouring into the dejected South had brought perhaps thousands of immigrants into Mexico within a few months. They had arrived, many with their families and little else, to find the choicest lands taken either by earlier settlers or speculators who hoped to hold on to the land until the price reached its zenith, then sell at colossal profits. In March Governor Allen pointed out some of the failings of the movement in one of his forthright editorials:

> Attracted by the promises of Mr. Maury spread through the United States, emigrants are arriving en masse, under the impression that disposable lands await them... Instead they find the lands all disposed of, and the proprietors each day becoming more exorbitant in their terms. The consequence is that all those who can do so return to the United States and take with them the most discouraging accounts…A prompt remedy is necessary, unless it is desirable to see the work of colonization, so happily conceived, miscarry.[17]

But remedies were not affected and discouragement began to take its toll. Matthew Maury sailed for England in March "on business" and to join his family with plans to bring them back to Mexico with him in October "after the rainy season." As a result of the developments of the next few months, he never returned.

Maury left his son Richard in charge of the immigration enterprise. Richard continued to receive requests for immigration information and free passage to Mexico. His replies to these inquiries also indicate the problems hampering the colonization program. In answering the request of a surveyor for employment early in April 1866, he mentioned that "agents of this office in the south have all been arrested, and required to give up their appointment; therefore no transportation can be furnished to you."[18] In another letter a few days later he elaborated on the situation:

> The Yankees have refused to permit our transportation agents to act in the United States and thus his majesty's intention of furnishing passage to needy emigrants has been frustrated. As immigrants

> are now coming faster than we want them, it is probable that this offer will not be renewed, as we can get as many colonists as we desire, without giving them this aid.[19]

Meanwhile those exiles already in Mexico continued with their plans for establishing new lives in their adopted country. In one of the smaller colonies, not far from Carlota, lived four expatriate families consisting of less than thirty people. They had taken residence in some dilapidated haciendas on the north bank of the Rio Blanco at Omealco, approximately twenty-five miles southeast of Cordoba. On the rugged south side of the Rio Blanco about four miles further down stream was an even smaller colony at Toloquillo. There a three hundred foot stone bridge spanned the river and rose at its arch some fifty feet above the stream.

Thomas J. Russell, one of the Omealco settlers, wrote in poetic terms of the area he had chosen as his new home.

> The beauty of this region can only be appreciated by being seen. From the plaza at Omealco looking westward, the Peak of Orizaba rises in its sublime majesty far above the clouds, with its white-robed summit of eternal snow, supported on either side by bold and rugged mountains gradually descending to the plain beneath. With a single glance of the eye you take in the whole range of terrestrial productions, from the groves of oranges, lemons, pine-apples, and coffee plantations, to the extreme limit of vegetable growth. Within the distance of three miles you can regale yourself with oranges fresh from the tree or pine-apples from the stock, and cool your mint julep with ice from the land of eternal frost.[20]

Though it might appear from this idyllic description that Russell attended the John Newman Edwards school of dramatic prose, the tone of his diary and his prospects for success at Omealco changed drastically with the sweep of coming events.

The Liberals throughout Mexico took heart in what they considered to be further manifestations of United States support and looked with increasing jealousy upon all foreigners, especially those Confederates who were settling on the choicest lands in Mexico. Among those Americans, unfortunately, were reckless men who had false notions about the honesty and morals of the natives and who were frequently the cause of serious offenses.

A group of Indians had contracted with some of the settlers in the Cordoba-Orizaba region to work their fields. After a few days' labor the Indians felt they had done enough work and refused to return to the farms. The colonists who went to their village to persuade them to fulfill the unexpired term of their contracts, lost patience with their excuses and forced them to return to their work. They had unwittingly furnished the local guerrilla chief an excuse to strike out at the *gringos.*

On the morning of May 15, 1866, "General" Luis Figueroa, claiming allegiance to Juarez and leading six hundred "soldiers" in the area some thirty miles north of Cordoba, surprised the sleeping settlers of the smaller colonies of Omealco and Tolequillo. After robbing them of everything they had of value, they marched them off as prisoners. Thomas Russell, one of the Omealco colonists, kept a diary of his captivity, which he carefully edited and later published in various newspapers and journals. He referred to Figueroa's "troops" as "cut throats and robbers who fought with Juarez mainly for the purpose of stealing."

The raid was probably meant to be punitive as Figueroa's men immediately announced that they were sequestering (taking and holding) the land on which the colony stood. The stunned colonists, now prisoners, stood by helplessly and watched the raiders burn and loot the town. Twenty-six of the colonists were rounded up and rudely interrogated. Figueroa flew into a rage when his men reported that a search of the town had turned up no pistols or rifles. The "general" then "threatened to hang every one of us to the nearest trees if we did not immediately inform him where we had hidden our arms, and made himself another search to see if any one had a pistol secreted about his person. On finding none, he ordered us into line across the plaza at a place where there was a high stone wall in our rear."[21]

Figueroa then called a company of infantry forward and positioned them facing the Americans standing against the stone wall. The Mexicans were obviously preparing to execute their prisoners. Just in time, Russell writes, "...one of the miserable wretches, who had taken a fine navy revolver from one of our men, came up to see the sort. The pistol was in his belt; the owner recognized it, and called the General's attention to it, giving the [serial] number." When Figueroa examined the pistol he found the colonist spoke the truth and, conducting a quick inspection of his own, discovered that his own men had confiscated all the weapons in the colony before Figueroa arrived on the scene.[22]

The captives were then led away and marched toward Vera Cruz and the sea. Left behind were three married men and their families and some infirm and old men who were cared for by the women in Figueroa's army. Though the Americans were treated somewhat humanely they were also insulted, spat upon and nearly starved as the bedraggled group stumbled, some of them shoeless, toward the Gulf of Mexico's shore.

Russell's diary tells how the prisoners evaded the firing squad for the first two days by relying upon their United States citizenship and producing passports to that effect. However, they had all become citizens of the Empire by virtue of settling the land set aside for colonization, a fact which came to light when one of Figueroa's men arrived in the Liberal camp with a copy of Maximilian's September Decree. The frightened colonists all had copies of the Decree as proof of their Mexican citizenship and suddenly realized that they were all possibly carrying their own death warrants. They were hard pressed to rid themselves of the documents, which they were finally able to do by discreetly burying or burning them. After satisfying Figueroa that they were not enemies of the Reform but only desired to settle the land peacefully and make a contribution to the progress of Mexico, the general told them that he had no objections to their living in Mexico as long as they had nothing to do with the Imperial Government.[23] He would shortly change his mind and issue a decree banishing them from the country or face death.

As the prisoners marched they passed through some of the most rugged and fantastic landscape in the Americas. On May 20 they arrived in a village of spectacular beauty, which, in spite of his ordeal, Russell described in inspiring terms:

> Jalapa, State of Oaxaca, May 20
>
> Arrived here today after a march of eight miles. This is but a small village of about thirty houses, situated at the base of a high cliff…To our right rises the mountain fully two thousand feet high, the greater part of the height being a perpendicular cliff of rock…The mountain range is actively volcanic, for we saw the fire and smoke coming out of it; and during the day, as we traveled along at the base of the cliff, smoke could be seen issuing out at several fissures in the rocks. The night following was dark and cloudy, and from our quarters in the village church we had a beautiful view of one of nature's sublimiest (sic) spectacles. Standing near the base of a tall perpendicular cliff which rose into the clouds, we could see the light from the internal fires illuming the heavens, waning and flashing and volumes of smoke that ascended in wreaths beyond the vale of clouds.[24]

The colonists struggled on for several more days. On some days they were fed, on others there was only water. By May 24, when they reached the Valley of the Rio Alvarada and marched through Paso Canoa, they were in fairly bad shape. Luckily, an American from New Orleans, Captain Thomas Soublet, who provided them with some food and medicine, met them at Tuxlepec.

Here Figueroa finally released most of his prisoners and sent them with Soublet to his nearby ranch; those not released were held pending further negotiations for their liberty. Little is known of Soublet other than that he was a rancher and entrepreneur in the region and apparently held sympathies for the Confederate refugees. He treated the colonists kindly and on May 30 arranged for some boats to take them to Cosamaloapam where he then negotiated further for passports so that they could leave the country.

Russell's diary entry for Saturday, June 2 says,

> Received our passports this morning, also a copy of the General's [Figueroa], degree banishing us from the country under penalty of death. Again we started on our way rejoicing. The distance is one hundred and fifty miles. Paddled our canoes nine leagues down river and then lay up for the night.[25]

When they arrived in Vera Cruz many of the survivors of the raid booked immediate passage back to the States. Others, including Tom Russell, accepted Soublet's arrangements for transportation to take them safely to Cordoba, where they hoped to find refuge. Russell's group reached Cordoba about 7:00 p.m. on Friday, June 12, 1866, to learn that their crops had been destroyed and the livestock that had not been stolen had been killed by "Liberals or robbers, it is not known which." No matter, in the estimation of many colonists there was little difference between the two.[26]

There apparently still existed a spirit of unity and compassion among the Confederate exiles. Later on that evening the Cordoba colonists threw an impromptu party at the Confederate Hotel for the survivors of the Omealco suffering.

The fate of some of the survivors of the Omealco raid is unknown. As for Tom Russell, he braved the death decree and remained in Cordoba for nearly four more months before leaving Mexico for good on October 1, 1866.

The Omealco raid had serious implications for the French as well as for the Confederates. If Liberal forces were in the area in such great strength, as was indicated by their bold attack on the colonists, that communications between Orizaba and the capital were in danger of being cut. Furthermore, the vital road from Vera Cruz passed through Paso del Macho a short distance from Cordoba. It could not be allowed to fall into the permanent possession of the Liberals.

To the Confederates it also meant that the small colonies surrounding Cordoba and Carlota were no longer safe. The colonists petitioned the Emperor for Imperial troops to occupy Omealco. In response, Marshal Bazaine sent a company of "contra-guerillas" which would be placed at Omealco "indefinitely" to protect the colony and nearby settlements. Measures were also taken under the sanction of the Imperial Government to increase the number of settlers for their mutual protection.[27]

As Russell's ordeal was playing out, the Empress Carlota departed Mexico on a mission to gain support in Europe for her husband and the Empire. She would not have entertained the thought that she would never see Mexico or Maximilian again.

On the very day she left for Europe, June 1, 1866, and almost to the hour, the revolutionary juggernaut swept over the Archduchess's namesake as hundreds of guerillas poured out of the jungle and destroyed the town of Carlota. The colonists there, as at Omealco, had to flee for their lives to escape capture. Many

Mexicans in the area who had collaborated with the colonists were taken prisoner and stoned to death. Several American prisoners were taken and tortured or shot down on their doorsteps. Most of the settlers managed to get away, but their exodus was bitter. Their trail toward the coast was strewn with suitcases, dead livestock, trunks containing possessions of all descriptions; and tragically, the bodies of several sick and weak colonists who were not strong enough to survive with the fleeing mob.

Following the Omealco and Carlota incidents other colonies were raided. "Six or eight peaceable and industrious colonists, representing various foreign nations," were taken; "some, in the fields, some from sick beds, and carried off with their hands tied."

It was reported that General Alejandro Garcia commanded a Liberal force in the area that numbered over 1,000 men, consisting of a large number of the workmen who had been laboring on the recently abandoned Imperial Railroad project.[28] It is probably more likely that the raids against the colonists were the work of mere robbers claiming to be Juaristas since Garcia had earlier publicly "disapproved in the strongest terms of this interruption to the labors of the colonists" and said that he "had always been in favor or colonization." He had also pronounced Figueroa a common bandit and threatened to hang him if he ever "got possession of him."[29]

Under Article 8 of the Immigration Decree the colonists had been given the right of forming "a stationary militia, for the purpose of protecting their property and neighborhoods." It is not certain whether such groups had been formed prior to the raids on the colonists, but if there had been such organizations, it is clear that they were not effective.

Perhaps the colonists did not take more aggressive action to protect themselves because of what Governor Allen called the American's propensity for "trusting too much in Providence." The Omealco incident did have one positive result, however; the settlers were frightened into action, and they began to form self-defense organizations. The *Mexican Times* was prompted to give the following advice:

> The Colonists should live in villages, where at least they should sleep every night, and they should cultivate their lands with their rifles hard by...They have now been taught a lesson by which if they are wise, they will profit...
>
> In addition...each separate settlement should have a military organization of some sort, and the means of communication with each other should always be at hand.[30]

The raids on the Cordoba settlements seemed to bring to the surface a hidden dissatisfaction that had been troubling the colonists for some time. There were scores of settlers in the larger settlements that could find neither land nor work. Others could not get the soil to produce due to a shortage of labor, a shortage of equipment, or a shortage of morale. Many of the later arrivals found that thieves and swindlers, who were as is their nature, attracted to such a high-minded enterprise, had preceded them. The adventurer, the dishonest man, the speculator and swindler robbed countrymen, the natives, everyone he could; then went home again "to practice morals in his fatherland."[31]

Since the United States had taken measures to discourage emigration the immigrant flood had dwindled to a trickle though there was still considerable interest throughout the South. Dysentery and "fever" took their toll during the rainy season because of the lack of proper medical care and perhaps, in the case of some of the poor, stranded would-be colonists, improper diet. Neither can discouragement and homesickness be discounted as a cause of the debilitating illnesses that began to plague the colonists in the spring and summer of 1866.

Governor Allen, the voice of colonization, had died during this period of illness and discontent, on April 29, 1866. In a letter he wrote near the end to family back home in Louisiana he expressed the nostalgia of many of his expatriate colleagues. "When I am sick in this lonely chamber," he wrote, "and I pass the hours and hours with no one but my Mexican servant to listen to the impatient ravings of a fevered brain, oh, then, I think of those dear ones I have left in Louisiana, of home and of all whom I loved so much."[32]

Henry Watkins Allen had come to Mexico because he could not bear to face a Federal military government in his native Louisiana, and though he tried to prepare immigrants for the hardships of colonization through the pages of the *Mexican Times*, he still encouraged them to come. He was buried in the American cemetery outside Mexico City in what was to have been a Confederate military burial. The United States Minister to Mexico intervened, informing the Confederates that they could not fly the Confederate flag in the American cemetery. They agreed and went on with the services, but it was rumored that Allen was buried in his Confederate uniform. Before his last illness he had already made clear his intentions to return to the United States and "resume his duties as a citizen, and urged many of his fellow exiles to do the same."[33]

P. H. Thompson, a friend of Governor Allen, had attended the funeral services and wrote his impressions to a friend in Louisiana:

> I was with him constantly during his last illness, and between the paroxysms of pain his mind constantly reverted to Louisiana, and his conversation was about the friends whom he had left there. The only wish he expressed was that he could once again see the waving fields of cane as they were in the olden time. Peace to his soul. We stood around his grave, that Monday evening in April, 1866 with the snowy peak of Popocatapetl towering above us, and as we listened to the sublime ritual few among that little band of exiles

> but thought of the homes which were once so peaceful and happy, and wondered wistfully would it be our fate to die away from our kindred, and without the gentle touch of woman on our brow.[34]

At the time this letter was written Thompson had every intention of remaining in Mexico, but all the colonists must have shared his almost sub-conscious concern for the future. They had hardly had time to reap even one of the two or three annual harvests Maury had promised them before interruptions by raiders and illness and other difficulties brought a pall of discouragement and gloom down upon the colonies.

Even before armed attacks on the colonists had begun William Marshall Anderson wrote of the worsening conditions in the colonies. With his characteristic concern for the people of Mexico, he noted the deteriorating situation in an entry in his diary dated May 18, 1866:

> Things are going to pieces here. God help these people. They would soon have been satisfied and contented, if it had not been for the noisy interference of the U. States. I would believe the bulk of the people and the best of the people would rather things should remain as they are than to change again. Where I have been, even in Coahuila, except [for] the greedy and speculative leaders, I firmly believe there are very many who would prefer a settled monarchy to an ever changeable and changing republic. They are weary and disgusted with this continued struggle and rapine.[35]

The situation continued to worsen throughout the winter and spring of 1866. Some, seeing the sign of the times, left before the decline became an avalanche. Thomas O. Moore, another former governor of Louisiana, returned to his plantation on the Red River in January, 1867, having fled from Mexico to Havana during the summer before.[36] Generals Preston, Smith, and John G. Walker went to Havana and one or two others planned to return immediately to the United States.[37] General John B. Magruder, who had fought the battle in the Office of Land Colonization through all kinds of disappointments and setbacks, wrote to a friend in Washington:

> I want permission to return to the United States. The Imperial Government has gone to hell, and has no money, and we have to get away from here. Many have already left; Price, myself and a few others remain, but we must look out for some place to go to. There is no hope for anything any more.[38]

Hardships, illness, isolation and constant attacks upon the colonies all combined to afflict the exiles with an outbreak of heartsickness. Making already intolerable matters worse, reports arrived on a regular basis reminding them that their fellow Americans back home felt little compassion for their travails. The *New York Tribune* carried an incredibly unsympathetic account of Carlota's destruction:

> The far-famed city of Carlota...consisting of a house, a barn and a stable, has been destroyed. The fields of coffee, by means of which Judge Perkins of Louisiana, expected to retrieve his lost fortunes have been ravaged; the pineapple plants, out of which General Price was to distill a most delicious fourth-proof brandy, have been uprooted...the palm shanties under the shelter of which about one hundred Southern emigrants have sought refuge have burned to the ground.[39]

Near the end of October, while he and his family were visiting friends in a nearby colony, Shelby's plantation was raided by local guerillas that destroyed much of his crops and property. In the mountains near Shelby's home was a rendezvous of bandits led by Manuel Rodriguez who had already made several raids in the area. Soon after the raid, Shelby, who was a sly tactician and strategist with years of experience in guerilla warfare before and during the Civil War, put his skills to good use by seeking a conference with Rodriguez. Meeting the Mexican chieftain in his own lair hidden away in the jungle was an extremely perilous mission especially for a "gringo." Nevertheless Shelby pulled it off, just how the truce was accomplished; no one knows. However, the raids suddenly stopped and for the next several weeks Rodriguez protected Shelby's plantation and others who had settled plantations near his.

Soon after the negotiations with Rodriguez, Shelby began to arm and train his hired hands to help defend the hacienda against further attacks. He went out of his way to treat his workers, and the locals fairly. He never forced any of the Mexicans or Indians who lived near or on his land to evacuate their homes as some of his fellow colonists had. Such arrogant and apathetic treatment of the locals by some Americans would be an aggravating factor in future attacks upon the colonies.

The Confederates also introduced a form of the American land system where a man with a family was allotted six hundred and forty acres at one dollar per acre. Men without families received three hundred and twenty acres at the same price. However, unscrupulous land speculators soon drove up prices and created even more hard feelings by selling land already occupied by Indians and Mexicans.

In that summer of 1866, came the high tide of discontent and discouragement. The Liberals became bolder in their activities, not only stepping up their harassment of the colonists, but also fighting larger scale battles with the French and Imperial Mexican armies. The unsettled conditions and dismal outlook forced prices down on land and commodities. The Mexicans were anxious to buy, but would rarely sell to the colonists, many of whom went wanting the necessities of life. Crops were destroyed and livestock stolen; then came the height of the rainy season bringing yellow fever and other illnesses. Those who had the means made their way to Vera Cruz to take passage for the United States. Those who could not secure anything beyond the bread on their tables stayed in their colonies and many degenerated into hopelessness and savagery.

> When their dead were lowered into the grave it was in silence, in haste, and without religious service. The amenities of life were scarcely recognized. The sick in many cases were neglected by countrymen and natives, and when the season of fever came on, and strong men dropped into the death-sleep in a day, the saddest sights were to be seen; scores there were of idle men in the streets, but none at the death-bed; crowds in the barrooms and at billiards, not at the burial. I speak of the mass; there were noble exceptions.[40]

The downfall of the Confederate colonization movement is actually two stories, for the decline of colonization accompanied the decline of Maximilian's Empire. Napoleon had tried to delay any decision that would take French troops out of Mexico, but American troops on the border and American supplies in the hands of the Liberals presaged a defeat of which the French Emperor wanted no part. He began to encourage Maximilian to abdicate, as he was afraid that if Maximilian remained in Mexico after the French troops were withdrawn and was forced to give up the throne, it would reflect badly upon France. Maximilian, beginning to see for the first time the motives of those who had invited him to take the throne, weighed abdication seriously but was discouraged by his wife, Carlota. The Empress decided to go to Paris to convince Napoleon to leave the troops in Mexico until her husband had a firmer hold. Though sympathetic, Napoleon could do nothing. An appeal to the Pope, who was indifferent since Maximilian had refused to return church lands, also failed, and the Empress, suffering from the tremendous physical and emotional strain of her self-imposed mission, began to go mad.

When Maximilian heard that his wife was insane, he drafted an abdication proclamation, shipped his baggage to Vera Cruz, and traveled as far as Orizaba where he spent the next six weeks. The Conservatives who made up the Emperor's circle of advisors still hoped to salvage some benefit for their cause and persuaded Maximilian to return to Mexico City and continue the fight. He decided in November 1866 to return to the capital, calling for another election, the results of which he would accept as the will of the people.[41]

In December 1867, Marshal Bazaine received instructions from Paris to begin the French withdrawal. All French soldiers, either serving within the Imperial Mexican army itself or with French units, were given the freedom to stay with Maximilian but only as members of the Imperial Army.[42] Bazaine made one last effort to persuade Maximilian to abdicate, and when Maximilian refused, he left the capital for Vera Cruz to oversee the evacuation of the army. In an attempt to force the Emperor to change his mind, the Marshal destroyed all the artillery and ammunition he could not take with him before leaving Mexico City.

News of the French withdrawal made it clear to all the remaining Confederate colonists that the quaking empire was drawing to an end. Without French troops the settlers could expect no protection at all from any source that could hold the Liberals and bandits at bay. It was clear to the colonists that their close identification with the Empire would not encourage friendly treatment from the Liberals. Even if Juarez should be disposed to extend protection to the colonies, it would be impossible since law and order in Mexico had traditionally extended barely beyond the limits of the largest towns. Hence the colonists stood alone, hated by the Mexican people and protected by neither party. A mad rush to quit Mexico ensued with homes being abandoned overnight. The former colonists crowded into the port city of Vera Cruz hoping to find space on the first available ship bound for the United States. Marshal Bazaine assisted many destitute colonists in obtaining passage, even paying their fare in some cases.

The *Mexican Times*, which had passed to John Newman Edwards' editorship upon the death of Henry Allen and was now published by B. C. Barksdale, tried to keep up an encouraging front as the Empire continued to shrink toward Mexico City. In January the *Mexican Times* carried an article that denied the abdication rumors and went on to recount another rumor that the editor had recently learned of "with pleasure." A scheme to settle the whole "Mexican difficulty" was under consideration which would allow the Emperor to sell to the United States a portion of Mexican territory extending from the Rio Grande on the east to Guaymas on the Gulf of California. The income from such a sale would be used to pay the French debt, and settle with other creditors, thus removing the question of foreign troops while the land transfer itself would remove the pressure from the United States. The basis for the story is unclear, and there is no evidence that such a plan was ever proposed to the United States; but that it should be printed at all illustrates the "dearth of news" Barksdale had complained of.[43] If true, the story might have helped to calm the worried colonists and restore confidence in the Empire.

As the Liberal forces became bolder with each departure of French troops, Barksdale gradually came to see France as the arch villain in the decline of the Empire:

> We hold that it is not against Maximilian or against Monarchy that the vast majority of Mexicans are now fighting, but against the immense French debt. Repudiate the whole French debt, and two thirds of the Liberal army would be disbanded at once...Repudiate this debt, renounce

> French claims, and the Liberal party would have no excuse or motive to fight against the Empire…
>
> And now, since Louis Napoleon has perfidiously, willfully and intentionally broken every contract and convention, and obligation entered into with Maximilian, and Mexico, shall the iron heel of his inexcusable interventional interference in the Mexican affair remain upon the radiant forehead of a people struggling for existence? [44]

The truth was that Napoleon had abandoned his Mexican venture not "perfidiously," but because he had no other choice. Popular opinion in France had turned against the intervention, France could not risk war with the United States, and Napoleon was faced with imminent war from Bismarck's Prussia that would require the employment of all his troops. He had tried to reach an agreement with some of the Liberal generals whereby Maximilian would be able to turn over the government to Porfirio Diaz, with an agreement that the new regime would pay French claims.

Such an arrangement would have saved face for all concerned by averting a humiliating defeat at the hands of Juarez. Diaz diplomatically declined the proposition, thus ending Napoleon's last hope of an honorable withdrawal. When Maximilian refused for the last time to abdicate, Napoleon could do nothing except write off the whole affair and withdraw his forces in the face Prussia's threat.

Bazaine departed from Mexico with the last French detachment, leaving Maximilian with an army of some 20,000 Mexican troops and a handful of European volunteers. Maximilian's Conservative advisors, hoping to avoid any further talk of abdication, convinced the Emperor that there was still hope of a military victory. They also convinced him that he was beloved of the Mexican people and that only he could bring justice, prosperity and peace to Mexico. These arguments and

Maximilian's own self-delusion persuaded him in February 1867 to take personal command of the Imperial Army at Queretaro. His fate was now on the wind.

The city of Queretaro was already under siege when Maximilian arrived, and every day more Juaristas joined the Liberal troops hammering the defenders. By the end of the battle for Queretaro the Imperialists were outnumbered five to one, with more Juaristas arriving by the day.

Maximilian finally paid the price for his indecision and for the faith he had placed in his advisors, when on May 14 one of his most trusted generals, Miguel Lopez, betrayed him to the enemy and, with no other way open, he was taken prisoner. Juarez also saw no other way but to place Maximilian before a military court martial to face Mexican judgment.

Maximilian's trial at Queretaro lasted well into June while The *Mexican Times*, now completely cut off from the news outside the capital city, continued to publish encouraging reports of his expected triumphant arrival by the first of July. "We have had unflinching faith in the Emperor and his army," read its last editorial. "We still hold to the old opinion that Maximilian still lives."[45]

The last issue of The *Mexican Times* was published on June 17, 1867. Two days later, on June 19, the Emperor was executed, having refused several offers of escape because he would not leave the two comrades, Miramon and Mejia, who were condemned with him. On June 21, General Diaz entered the City of Mexico. The *Mexican Times* was finished; the "voice of colonization" had ceased to speak.

CHAPTER EIGHT

The Adventures of Jo Shelby in the Cordoba Valley

Although Shelby was no longer a soldier he would need all his martial skills to survive as a colonist in the Cordoba Valley. Cordoba was an old Mexican town and the focal point for a cluster of Confederate colonies scattered around the Cordoba Valley.

Two miles outside the town of Cordoba stood the stone, Spanish-tile roofed hacienda Maximilian sold Shelby when he first came there. It was said that the grand hacienda had previously belonged to General Antonio Lopez de Santa Anna who was at the time living in exile. It was Santa Anna who had suspended the Mexican Constitution in 1835 and then led an army into Texas to put down the rebellion that resulted in Texas independence in 1836. Santa Anna was a man who enjoyed the finer things of life, and, while there are no contemporary descriptions of Shelby's new home, it may be assumed that it must have been an extraordinary dwelling.

Shelby immediately went to work and began to carve out a new life for himself and his family who would soon join him. He was an experienced farmer and businessman, and immediately recognized the potential of the agricultural paradise he had settled in. He began with twelve acres of coffee and an aggressive renovation of the plantation. By mid-September he had the operation up and running.

In early October 1865, Elizabeth (Betsy) Shelby boarded a river steamer at Lexington with their two sons, Orville and Joseph. Around her waist she wore a money belt containing five hundred dollars in gold. After a stop at St. Louis the steamer sailed on to New Orleans where Betsy and the boys boarded a ship bound for Vera Cruz, Mexico, arriving there on October 14, 1865.[1]

Shelby had not seen his family for over two years and their meeting on a foreign shore must have seemed a bewildering experience. With so much time and tumult between them through those years it was difficult to become immediately reacquainted but there was little time to do so gradually. They must move into their new home and build a new life or face all the negative affects of failure.

Shelby had personally arranged a truce with local bandit leader Manuel Rodriguez, which gave his business interests some breathing room, but life at Cordoba presented other problems. By the end of October 1865, he joined with Major McMurty, a confederate officer whose more complete identity is lost to history, in a freighting business operating out of Cordoba. Using ten mule wagon teams running between Paso del Macho and Mexico City, their new freight line quickly expanded, and the operation soon became a thriving business when Shelby received a contract to supply the French military garrisons along the rail line. "Strategically, as well as ideologically," this was a perilous position for them because the French were perpetually under attack from the Juaristas.[2]

Shelby must have realized the danger his freighting business was creating when he began to supply the French outposts along the railway, and, although he was still unsure regarding his future in Mexico, there was one thing he was certain about: he did not want to return to the United States under any circumstances. He believed many of his fellow countrymen had sold out their cause because they loved the comforts of the old slave-supported life more than they did the Confederacy. In a letter written to a friend, Frank Lilly, Shelby's bitterness at those who had surrendered to the Yankees in Missouri and Kentucky to save their properties from destruction is demonstrated in this fire and brimstone tirade:

> I am here as an exile; defeated by the acts of the southern people...Let them reap what they deserved, eternal disgrace. D—n' em, they were foolish enough to think by laying down their arms they would enjoy all the rights they once had...My heart is heavy at the thought of being separated from you all forever; but I am not one of those to ask forgiveness for that which I believe today is right. The party in power has manifested no leniency.[3]

With the South at the mercy of the radicals it would have been insufferable to him to be forced to take a humiliating, and in some cases, ridiculous oath of allegiance. In Missouri that oath was more severe than the normal federal test requirements. Many of the exiles in Mexico had come from Missouri, and to hold any sort of public office again in their home state they would be forced to swear their innocence on eighty-six separate issues. Further they must swear that they had never acted in a hostile manner toward the government of Missouri. To men like Shelby who nursed an exaggerated sense of honor such oaths were anathema.

Southerners who were hoping for a presidential pardon had to take the "iron clad" test and oath of allegiance with the knowledge that if they were denied a pardon there was a distinct possibility of being jailed. Henry Winter Davis and Charles Sumner hoped to deprive the former Confederates, all the exiles who went to Mexico, of their full citizenship forever. There was a very real possibility that Shelby and others could be charged with treason and imprisoned, and in a worse case scenario, even perhaps face the gallows.

The amnesty proclamation issued by Andrew Johnson showed his own bias toward the wealthy, land-owning class of the former rebel states. This hated decree required that any Confederates who owned property worth more than $20,000 had to have their cases reviewed by the Union authorities and their pardons must come directly from the president. Ironically, there were few men in the south who had fought for the Confederacy who had anything left to come home to after the war, much less $20,000 worth of land and property.

Shelby may have been the hardest of hard cases but his defiant attitude toward the offers of pardon was typical among the Confederate exiles. He was not willing to compromise his honor by dealing with the pardon-brokers who were arbitrarily handing out or denying amnesty on what seemed to him to be faithless whims. There were also other reasons to be apprehensive about returning to the United States considering the many Confederates who had been captured and imprisoned during this time.

Maintaining powerful feelings of defiance toward Yankee supremacy, many exiles like Shelby were disappointed in Andrew Johnson's evident detachment concerning their special circumstances under the new amnesty rules. One of these rules excluded from amnesty anyone who had left his home in the United States to aid the Confederacy. Johnson, a fellow Southerner and Democrat, could not, or would not, persuade the United States Congress to allow an exception for the exiles on this point. There was more than enough ambiguity in Johnson's attitude to make

Shelby and the other exiles wary of trusting the Federal government in this matter.

Former Missouri Governor Thomas Caute Reynolds shared Shelby's views in a letter he wrote from Mexico to Jubal Early.

> I quietly await the results of the contest [Reconstruction], consider myself still a citizen and resident of Missouri, and hope and expect, God willing and life lasting, to return to it, as one of the ruling class. Meanwhile I wish success to all efforts to restore the South to equality and power; towards such as wish to oppress or injure Confederates, my attitude personally is simply one of calm, contemptuous, derisive defiance. The whole nation, from the President down, with all its fleets and armies, is powerless to hurt a hair of my head; and, if they think at all of it they know it.[4]

Ironically, they [the radical unionists] never did "hurt a hair" of Reynolds head; later circumstances would leave that to Reynolds himself. Despite his defiant stance against the United States Government Reynolds did return to Missouri following the collapse of the Mexican colonies, but he was never able to completely recover from the trauma and scenes of horror he had witnessed in Mexico. He also harbored personal feelings of shame and regret for the advice he had given Maximilian, which he believed had contributed to the Emperor's execution. For a time he entertained thoughts of recapturing a high public office. Eventually he was elected to the Missouri General Assembly in 1874, but later fell into a deep depression and drifted aimlessly through one business venture after another. On March 30 1887, Reynolds committed suicide. He climbed the stairs to the top of one of the tallest buildings in St. Louis and then walked backwards "three steps" into an empty elevator shaft and fell to his death.[5]

While Shelby's plantation and freight business gained momentum, he and his wife awaited the birth of their third child, the only one of their eventual eight sons to be born in Mexico. However, if he thought he had found a lasting peace in his new country and that his days of fighting and killing were over, he was mistaken.

In February 1866, Shelby was escorting eighty wagons filled with supplies for French Colonel DuPrueil when he was attacked by Juarista guerillas just north of Mexico City. Among the teamsters that day were several of his troops from the Iron Brigade. Heavily outnumbered the Confederates and other teamsters were forced to abandon their wagons, mount up, and shoot their way out of the ambush eventually reaching the French lines near Mexico City. The monetary loss from this ambush was significant for the company and probably made him realize that his dreams and plans for a new life were in peril.

Later in March 1866, Shelby and an escort of his former Confederate cavalrymen came under attack once again as they attempted to re-supply Colonel Duprueil at his isolated command at Censola where he was facing a superior force of Juarista rebels. General Felix C. Douay, commanding the French troops at San Luis Potosi, sent Shelby and twenty men to escort ten wagons loaded with supplies and ammunition into the no man's land north of the city in an effort to reinforce the besieged French outpost.

DuPreuil commanded six hundred men of the Foreign Legion at the post "twenty leagues" northwest of San Luis Potosi. Douay, stationed there, had an inadequate cavalry force and was barely holding on while awaiting his own reinforcements from Marshal Bazaine in Mexico City. The guerillas "some two thousand strong" and "emboldened by the absence of cavalry had risen up" and attacked the French troops at Censola.[6] Shelby's mission was to make his way through the Mexican forces swarming over the countryside and attempt to reach DuPreuil before his command was overrun.

The irony of the situation was curious. The first time Shelby and DuPreuil met they nearly fought a duel. This time, however, the French commander was dependent on the former major general of the Missouri Cavalry Division for his own survival. For General Douay and Colonel DuPreuil the circumstances of this incident were indicative of how quickly the fortunes of the French forces had deteriorated all over Mexico. The Archduchess Carlota and the Emperor had done everything in their power to keep law and order and to protect the civilians from the guerillas; but the end was coming soon for the French Empire in Mexico.

Among the twenty men Shelby led out from San Luis Potosi were Captain James Kirtley, James Wood, Sandy Jones, Thomas Boswell and George Hall. All were seasoned combat veterans who had been fired upon in dozens of battles during the war and from Missouri to the Gulf of Mexico. Many of the other teamsters were volunteer colonists and Shelby's Mexican and Indian laborers. Over the past two months Shelby and his veterans had put the less experienced men through a crash-training course in firearms and basic military tactics for use in close quarter engagements.

Their objective, Censola, lay approximately sixty miles northwest of San Luis Potosi. Shelby's party had traveled nearly half that distance when some five hundred Mexicans led by renegade priest Ramon Gutierrez ambushed them. Their first volley seriously wounded James Wood and Sandy Jones. Shelby recognized immediately that he was at a disadvantage and ordered an immediate counter march to a local hacienda where he intended to fortify a position between the house and the corral.

When the teamsters reached the hacienda they frantically constructed breastworks with overturned wagons, bags of flour, beans and rice bound for DuPreuil's command, and several hastily filled burlap sandbags. Shelby's men took up defensive positions and immediately prepared to give battle. Captain James Kirtley and two other men were ordered to "ride like the devil" back to San Luis Potosi, report to General Douay and ask for reinforcements. To reach Douay, Kirtley's riders would have to cover nearly thirty miles of rough terrain through hundreds of guerillas roaming the countryside. Until help arrived Shelby would attempt to hold his position against five hundred enemy soldiers with only seventeen men.

Soon after Kirtley's party rode off, the Juaristas attempted to rush Shelby's position, probably hoping for a quick and easy victory. A barrage of rifle fire drove them back, leaving several dead bodies behind. Pushed back out of rifle range the Mexican guerillas began to regroup. Nightfall was only a few hours away and Shelby realized that when darkness came he would not be able to prevent the enemy from flanking his position and surrounding his small force.

He had his men cut and strip the bark from "two dark looking logs," then had the logs coated with a layer of axle grease. Once mounted on makeshift wagon wheels the greased logs were placed at each end of the barricades; "at either angle of the corral…these logs, from a distance looked like cannon."[7] Shelby's wooden-barreled gun batteries were a stroke of improvisational battlefield genius. For the rest of the afternoon, the fake guns would be a determining factor in causing Gutierrez to hesitate and think twice before throwing an all out assault against Shelby's position.

Shortly after nightfall, Kirtley, Thomas Boswell and George Hall were pushing their animals on through the pitch-blackness as hard as they dared. They knew Shelby might be fighting for his life or, for all they knew he and his men might already be dead. Thus motivated they continued the reckless pace.

Meanwhile, Shelby's men waited behind their barricades, wondering if the Mexicans would mount a night attack. An occasional round fired from somewhere out in the darkness thudded into the bags of beans and rice and sand. It was going to be a long night for the Americans who prayed quietly that Kirtley and his companions had gotten through to San Luis Potosi.

Just before daylight was breaking, Shelby's scouts rode their worn out horses into San Luis Potosi. After briefing General

Douay, and with no sleep or rest, they were ready to mount fresh horses and head back to Shelby's rescue.

> Before the sun rose three squadrons of Chasseurs, a section of flying artillery, and the Americans thrown forward as guides, were galloping back towards the hacienda at which Shelby was fortified and fighting an overwhelming army of guerillas.[8]

As the sun rose over the barricades that same morning, Gutierrez sent a message to Shelby demanding his surrender. He refused, but in hopes of buying time requested a meeting with the Mexican leader to discuss terms. He and the rebel chieftain met some one hundred and fifty yards outside his works, from where his adversary was unable to get a clear look at the fake gun batteries. During the negotiations Shelby feigned possible surrender to buy even more time. Before returning to his lines, he mentioned to Gutierrez that he did not think the bandit chief actually had enough troops to force him to surrender. Gutierrez promised to give Shelby a demonstration that would change his mind. The Mexican still did not want to chance a frontal assault against two gun emplacements. If possible he preferred intimidating Shelby into surrendering before he was forced to order a charge.

Once Shelby was back behind his barricades, Gutierrez began to display his troops at various points and to march them about to prove his numerical strength. Another hour was lost to the Mexicans as Gutierrez flexed his muscles in hopes of scaring Shelby into submission.

The real problem for Gutierrez was that Shelby had never intended to surrender and precious time was passing. Once the lengthy display of martial strength was terminated, Gutierrez sent in his final message demanding surrender. Shelby refused, of course, but he did so knowing full well, as John Newman Edwards put it, "that 500 sheep might be able to run over 17 men," let alone five hundred soldiers. Nevertheless, Shelby had played out his string as long as he could and now time was up. He had no idea if Kirtley had gotten through to Douay or, if he had, how close his rescuers might be.

The Mexicans began to form their battle lines with some one hundred of them mounted and posted on the flanks. The Confederates braced themselves to make their final stand. A born and bred cavalryman, most surely one of Shelby's reputation, did not want to make his last fight confined behind breastworks. What might he have given for a company of his troopers from the old Iron Brigade to sally forth with him on horseback to meet the enemy?

Suddenly, the Juaristas rushed their position, firing as they came.

> With a shrill, short yell the Mexicans dashed forward to the attack. Had the wave held its course it would have inundated the earthwork. It broke, however, before it reached half way across the open space behind which it had gathered for the onset. Those in front began to fire too soon, and those in the rear, not seeing from the smoke what was really in front, fired too, and without aim or object. With unloaded guns they dared not go on—the fire of the Americans was distressing beyond endurance—the wave broke itself into fragments—and the sun sunk lower and lower.[9]

When the charge faltered, Shelby began exhorting his men to hold fast. "We're nearly out of the wilderness boys," he shouted.

One of his troopers standing nearby was showing the strain of the hours of combat. His face was blackened from powder smoke and his parched lips were beginning to crack. "We need water greatly, General," he said.

Shelby gave the man an encouraging look and replied, "In an hour you shall drink your fill…for in an hour the French will be here."

Another man shouted, "But what if Kirtley has fallen?"

Their commander's reply came from a man who had been in many deadly confrontations that had taught him which soldiers he could depend upon in crisis. "He will not fall," Shelby called back. "Luck goes with him everywhere."

Just then, as though John Ford of the movies had scripted it himself, the cavalry literally came to the rescue. Gutierrez' men were preparing to make another assault when suddenly, "the full, ringing notes of the French bugles were heard, and in a moment more the squadrons emerged from the trees, galloping straight and in beautiful order towards the guerillas."[10]

Within short order the Juaristas had deserted the field and fled. Captain Mesillon of the relieving force sought Shelby out and handed him a dispatch written by General Douay. It was a most amazing order considering the inherent belief among the French military in their own superiority. By order of General Douay, Shelby, an American, was placed in command of a company of French cavalry with the same authority as a regular French officer commanding troops in the field. The order also gave Shelby a new mission concerning DuPreuil's situation at Censola. General Douay's message read:

> To complete the conquest of Colonel DuPreuil, of whose bearing towards you at Parras I have been duly informed by Gen. Jeanningros, I choose that he shall owe his life to you. Capt. Mesillon awaits your orders. I need not advise you to be circumspect, and to take your own time and way to reach Censola and bring my Frenchmen back to me, for whom, I imagine, there is no great love in the hearts of its inhabitants."[11]

Placing himself at the head of the company of French Cuirassiers, Shelby explained to the young Captain Mesillon that the plans had changed. They were no longer planning to reinforce the French outpost, rather they were going to drive forward into Censola and then lead what was left of DuPreuil's legionnaires out of the besieged town and escort them back to San Luis Potosi.

To make his new French command more mobile Shelby left his wagons behind and started toward Censola knowing full well that a month earlier, on February 12, 1866, Secretary Seward had demanded that the French withdraw from Mexico or face military intervention from the United States. Soon after, Napoleon duly ordered Marshal Bazaine to begin preparations to pull the army out of Mexico and return to France where they would be needed in conflict with Prussia, which was most certainly coming.

For Shelby and the Colonies it was almost over as soon as it had begun. Once Napoleon made his decision, the Mexican enterprise began to deteriorate with startling rapidity for both the French and the Confederate exiles.

As there were no Imperial troops in the entire area Shelby's French detachment was riding into a darkness of uncertainty. If they were to be cut off or pinned down, there would be no possibility this time of reinforcement or rescue. They would either have to cut their way out or fight to the last man.

When night fell Shelby's column continued to push hard toward Censola. On seven separate occasions during the ride they were bushwhacked by Juarista troops and guerillas. Before they reached Censola just before sunrise, twenty-eight French cavalrymen had been killed in these nocturnal attacks. DuPreuil's legionnaires were extremely relieved, and surprised, to see Shelby at the head of the French relief column.

The two former antagonists greeted each other like old friends and immediately set to making preparations to get the hell out of Censola before the Juaristas could mount an assault on the town; or cut off their route of escape. Within a few hours DuPreuil's command had geared up pack animals and with gun batteries in tow were ready to march.

> Still leading, Shelby came away from Censola in command of the whole French force. DuPreuil's men wondered a little, but DuPreuil, in the height of his gratitude, thought no compliment sufficiently high to pay the rough-clad, quiet American fighter, who did not even so much as have a red sash around him as insignia of rank or authority.[12]

Only a few miles from Censola the fighting commenced anew. Attacks and ambushes continued throughout the next two days and nights. And each time Shelby made his attackers pay dearly. He countered each attack with

> several sudden and bloody repulses to put such a wholesome fear of punishment on the minds of the pursuers that they gave him ample time to carve out for the train a safe road in front while protecting amply the perilous road in the rear.[13]

On more than one occasion Shelby led the French cavalry in charges that overran or dispersed the enemy. The French cavalryman's weapon of choice was the saber, as it had been for the Union cavalry during the first three years of the Civil War. Shelby taught the French under his command a new reality by demonstrating the deadly effectiveness and advantage of using two Navy six-shooters in close quarter combat.

For three days and three nights Shelby held steady to his course, fighting constantly and caring for his dead and wounded. On the morning of the fourth day he led the bedraggled column through the French lines outside of San Luis Potosi and into the city where they were greeted by the entire garrison which was turned out to cheer for their returning comrades. Douay poured congratulations and thanks on his new American commander and demanded to pay for his losses in mules and wagons.

"Very well, General," Shelby replied, "but as you are leaving the country you must wait until we meet again in the City of Mexico. Until then, remember your promise."[14]

Judging from all Shelby's experienced eye had seen he knew that for him and Douay there would not be another meeting. The garrison at San Luis Potosi was one of the first to be evacuated. Soon, the European troops would begin a general withdrawal, but for the time being the highways of Mexico became a bandit's paradise, rendering the freighting business no longer feasible.

At heart Shelby had always been a farmer and now he returned to the idea of establishing one or more plantations in another area of Mexico. The recent war had taught him the value of adaptability, so with the freighting enterprise falling apart on him, Shelby now became interested in a colonization project at Tuxpan, where Baron Enrique Sauvage, a French financier had acquired a huge tract of land the size of Delaware.

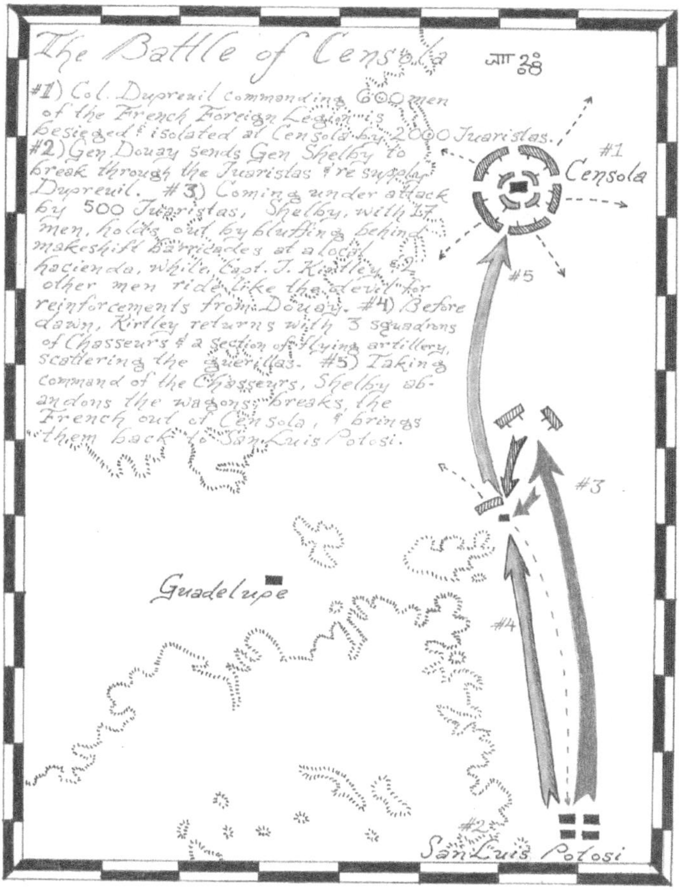
The Battle of Censola
JM 28
#1) Col. Dupreuil commanding 600 men of the French Foreign Legion is besieged & isolated at Censola by 2000 Juaristas. #2) Gen Douay sends Gen Shelby to break through the Juaristas & re supply Dupreuil. #3) Coming under attack by 500 Juaristas, Shelby, with 17 men, holds out by bluffing behind makeshift barricades at a local hacienda, while Capt. J. Kirtley & 2 other men ride like the devil for reinforcements from Douay. #4) Before dawn, Kirtley returns with 3 squadrons of Chasseurs & a section of flying artillery, scattering the guerillas. #5) Taking command of the Chasseurs, Shelby abandons the wagons, breaks the French out of Censola, & brings them back to San Luis Potosi.
#1
Censola
#5
#3
Guadelupe
#4
#2
San Luis Potosi

CHAPTER NINE

Tuxpan, Shelby's Last Stand

So difficult had been their trials that by the end of summer, 1866, only a few exiles braved the new conditions in Mexico under Benito Juarez. Among them was Jo Shelby, still optimistically seeking his place in this country of his exile. Baron Sauvage dreamed of great wealth from his Tuxpan project and Shelby caught the fever, albeit with obvious reservations because of the catastrophes at Omealco and Carlota. But the new promoters believed their chances of success at Tuxpan were better due to the remoteness of the new location.

The anticipated Tuxpan project lay a hundred and forty-five miles north of the Cordoba colonies. The new colony would be located some twenty miles inland from the mouth of the Tuxpan River where it flows into the Gulf of Mexico between Tampico and Vera Cruz. Near there also resided the Toluca Indians, a primitive tribe usually considered to be friendly. They claimed some 10,000 acres, which were perfectly suited for coffee, sugar cane, corn, and a variety of fruits.

This area was also considered one of the most fertile areas in Mexico, covered in mahogany and rubber trees and rich in mineral wealth. However, its greatest wealth, yet to be discovered, were the most prolific oil deposits in the Western Hemisphere. These oil riches lay undiscovered just a few hundred feet beneath the surface of the land Shelby saw as an agricultural paradise. Had the Americans buying these tracts of land been able to look into the future and see the power and wealth that oil would bring they might never have relinquished their land rights come what may.

A descendant of one of the southern families who had attempted settlement in Tuxpan returned to Mexico in 1929, following the discovery of the great deposits of black gold. He spent some time in the Mexican capital attempting to secure legal validation to a land claim reaching back to the Maximilian era.[1] Apparently his efforts bore no more fruit than the sweat and blood of his ancestor.

On October 8, 1866, the *Mexican Times* reported that Maximilian had signed a decree permitting Baron Sauvage to transfer his title rights in Tuxpan to Jo Shelby in return for stock in the new company.

> Every emigrant will be landed free of cost within two days march of his settlement; agents of known energy and discretion are to be sent through North America, and the stock of the company will be offered at once in the New York market. The lands are government, without being obtained by confiscation, and are among the richest in Mexico. It is impossible to overestimate the great advantages of the colony located near Tuxpan will be to the government and the friends of improvement and agricultural wealth, and the great necessity for the

> immediate pacification of this department is most evident. The destiny of the Government is in its own hands now. Colonization, unfortunately mismanaged at first, comes back again to the vision on the horizon of Mexico, and when Shelby and Laurence Keith Boyle and others of similar energy and intelligence put their shoulders to the wheel success is easy and colonization already commenced. [2]

Laurence Boyle was an American entrepreneur in Mexico City who soon traveled to New York to seek backing from Wall Street. On November 19, 1866 John Newman Edwards reported for the *Mexican Times*:

> We conscientiously believe that had the management of affairs at Cordoba been trusted to his [Shelby's] keeping today a great and growing colony would have been stretching browned hands southward from the Gulf and westward to the gates of Mexico, holding up the white beacon of cotton bales and beckoning from the world commerce, enterprise and capital. With him there is no such word as fail, and the man who went up from captain to major general along a path illuminated only by the light of his sword and the blaze of his genius, will find it easier to destroy the obstacles of nature alone than to overcome the barriers of man and nature combined, as he always did. When all arrangements have been perfected we will have other and more extended information to our readers. [3]

It is difficult to decide which was more outrageous: Edwards' hero worship of his old commander, or his optimism for the new colony at Tuxpan; especially considering that the *Mexican Times* would soon be going out of publication permanently.

Former Confederate Engineer, Major R.J. Lawrence from Kansas City, surveyed the land at Tuxpan and Maximilian's treasury kicked in a subsidy of $20,000 to help build a railroad from Tampico to Vera Cruz. Shelby's colonization program attracted wide attention, and soon Shelby had two hundred Mexican laborers at work at the new colony. He acquired a new home in Tuxpan for Betsy and the children. About the same time he purchased a seaworthy schooner that he soon sailed to Havana, Cuba, where he hired a new crew and then sailed back to Vera Cruz with a load of supplies. At Vera Cruz Shelby bought two more ships from the French Coast Guard and then, hearing that Bazaine was also in Vera Cruz at the time, he went to pay the Marshal a visit.

Shelby wanted to make certain that he did all in his power to provide security for the colony at Tuxpan should the worst come. This was foremost on his mind when he found Bazaine, and he wasted no time coming directly to the point:

> Marshal[,] we have taken upon our hands much work. We have plenty of farming implements but no guns. I understand that your army of occupation has been supplied recently with some new Chassepot rifles, and I am wondering if you couldn't spare us some of your old muskets. With 500 muskets and enough ammunition to protect the colony at Tuxpan for six months, I promise you that we can build a colony as prosperous as the colony of the Emperor Napoleon in North Africa.[4]

Bazaine seemed amused by the tenacious American, but immediately signed the order for five hundred muskets and thousands of rounds of ammunition. Shelby loaded his newly acquired arms aboard one of his vessels and sailed out of Vera Cruz across the Bay of Campeche, feeling confident that, with a little luck, he could make a commercial success of this last stand at Tuxpan.

From November 1866 to January 1867 Shelby drove himself like a man possessed. All at the same time he busied himself working on a new railroad line, operating his new plantation, and overseeing the development of the colony project. In addition he also began a schooner freighting business carrying coffee, beans, rice, sugarcane and tropical fruit between Tampico and New Orleans and even joined a venture with Baron Sauvage and Richard L. Maury (Commodore Maury's son) to lay an undersea telegraph cable from Vera Cruz to Cuba.

Shelby had less than a year to get things established in Tuxpan before the return of Benito Juarez to power would force him to consider what, until then, had been unthinkable: return to an uncertain future in Missouri. The exact date of the first attack on Tuxpan is unrecorded. The best estimates place the date somewhere in the first or second week of February 1867. However, Shelby had been fighting small skirmishes with the Indians and guerillas for several months before the final blows were delivered. And when the French garrison at Correzetla was evacuated and sent back to Mexico City, the colonists at Tuxpan were left without any protection from the Imperial army.

The colony's dismal situation did not go unnoticed by their enemies. The Tolucas, urged on by the Juaristas, began to demand the return of their mahogany forests and other lands claimed by the colonists. In the weeks since the Imperial troops had been withdrawn to Mexico City, Shelby had been leading a small contingent of colonists and loyal Mexican workers against the Tolucas in bloody hand-to-hand jungle fighting. Soon the Toluca insurrection had grown to a two-thousand-man army commanded by General Porfirio Diaz.

The volunteer militia of Americans and Mexicans managed to drive the guerillas back for a time. But when Shelby's scouts reported that thousands of Indians and Mexicans were moving on Tuxpan through the mahogany forest chanting, "Death to the Gringos," he was forced to start an immediate evacuation of the colony. He knew that with their backs to the sea they could very well be overrun and slaughtered if they stayed.

Leaving virtually everything they owned except what they could carry on their backs, some one hundred colonists, men, women and children, began their desperate march towards Vera Cruz and Cordoba. As he had done numerous times during the war, Shelby fell back with approximately twenty men to provide a rear guard for the fleeing exiles. The lead elements of the Tolucan army swept through Tuxpan burning homes and killing any stragglers and other colonists they came across in the outlying areas. Shelby's men fought ferociously to hold off attacking parties of Indians as the column of refugees put more ground between them and the carnage occurring at their abandoned and burning homesteads.

The Indians had no mercy on those colonists cut off and trapped in Tuxpan. In isolated pockets of resistance some men and women fought to the death on the beaches with their backs to the Bay of Campeche. Some boarded small rowboats and made for the schooners anchored in the harbor at the mouth of the Tuxpan River while a group of them dug ditches and threw up earthworks to hold off the enemy. In short order the Indians overwhelmed the men fighting from their shallow trenches.

As the survivors swam for the schooners or furiously paddled what few boats they had, the Indians gave pursuit in boats of their own, killing them on the water then boarding and setting fire to the ships. To avoid the flames some of the colonists, whose clothing had caught fire, jumped into the ocean and drowned. Others were caught and tortured. This "Garden of Eden," to which they thought they had come only a few short months before had in the blink of an eye become a hell on earth.

When Shelby and the main column of refugees from Tuxpan arrived at Tampico a disappointing and alarming scene greeted them. Shelby's two ships that he had left harbored there had been scuttled and their crews murdered. There was nothing they could do but continue their march toward Vera Cruz where they hoped to find sanctuary.

They reached the city in the first week of March 1867. Shelby and a few others continued on to Cordoba, arriving there a few days later. He found the Hacienda de Santa Ana still intact, but with the fall of Maximilian's government his title to the land was no longer valid. The situation here in Cordoba, as in all the besieged colonies, was extremely dangerous. Though Shelby had the grit to face what came with the tenacity and courage he had displayed in dozens of battles, he had the innocent to think of. He turned to his faithful adjutant, John Newman Edwards, recently editor of the *Mexican Times* until Mexico City had fallen to the Juaristas, to escort his wife and three sons back to New Orleans and thence to St. Louis.

Before leaving, Betsy Shelby divided her jewelry up between Shelby's men, keeping just enough for her passage to St. Louis. "When they boarded the ship at Vera Cruz all they took with them from Mexico was a little dog which Archbishop Labastida, Archbishop of Mexico, had given the Shelby children."[5]

Jo Shelby, himself, planned to remain in Cordoba for a few more weeks to see what could be salvaged from the wreckage of the colony. If he survived he would later join his family in Kentucky.

A bitter specter was beginning to stalk Shelby. From Cordoba, he wrote to his friend, Major R.J. Lawrence, "One thing is certain, we must all get away from here or I will be damned if we don't starve--there is no joke about it." Shelby had heard that Sterling Price had abandoned the colony and left for the states. The rumor also said that fifty thousand dollars was being raised for the old soldier's relief. Shelby's disgust at this was reflected in the closing lines of his letter to Lawrence: "They had better damn sight appropriate that fifty thousand to the orphans and widows that were made by his damn blunders."[6]

Not all of the ex-Confederates who went to the Cordoba Valley were as young or as iron-willed as Shelby, who could be extremely judgmental concerning his fellow officers. He had tried hard to keep under control his critical feelings toward those who had surrendered the Army of the Trans-Mississippi. What little respect he had left for the competence of his peers, was shattered completely when Price and others, in his eyes, gave up and betrayed him once again.

After safely escorting Elizabeth and her children to New Orleans, Major Edwards held an interview with the press concerning his own feelings on Mexico.

> There is no love for the people of the United States, and the only sympathy for them at all is because they are presumed to represent opposition to the French and to have expressed dislike for them. The settlement at Cordoba is among things of the past. The departure of General Price, which has been duly chronicled, will be followed by almost all who associated with him at Cordoba. Judge Perkins has gone to Paris, France. Governor Harris left two weeks ago for Havana. General Shelby still remains in Cordoba, and probably will remain for several months to come. Condition of the country is worse than has been known for twenty years. The main thoroughfare between the city of Mexico and Vera Cruz is interrupted about every fifteen miles and everything the lucky passengers possess is taken with a quiet shrug of the shoulders and the polite declaration: *no me importa*, which means that it makes no matter to me.[7]

Shelby finally returned home during the second week in June 1867 aboard the United States gunboat *Tacony*. The Navy Department had sent the gunboat to Vera Cruz to protect American citizens after the French army had left Mexico. He was prepared for the worst, but upon his return from exile, as things turned out, he was well received; and in many cases, lionized.

In New Orleans Shelby boarded the *Great Republic*, a steamer bound for St. Louis. From there he took the last leg of his journey by rail to Lafayette County, Kentucky, where his family awaited his return. Soon he was farming several acres of hemp near Lexington and began managing a coal mining operation in Clarksburg, Missouri. In 1869 Shelby became a financier and general superintendent for the Lexington and Sedalia Railroad, which he hoped to link up with the new transcontinental railroad that was about to change the face of America by connecting the west and the newly acquired Pacific ports with the rest of the nation.

Among the varied visitors to the Shelby homestead in those years were some of his old friends from the war and men he had ridden with during the bloody border wars in the late 1850s. On numerous occasions Frank and Jesse James, along with Cole Younger and his brothers, dropped by to visit or receive help. Shelby never turned his back on his former comrades in arms, even if at that time they were the most wanted outlaws in the region. Because he was one of the most famous and influential men in Missouri, the former major general decided to brave the controversy and political risk of his association with known outlaws, and in 1893, still loyal to old comrades, testified for the defense at the trial of Frank James.

Apparently no negative consequences of his association with outlaws accrued, for that same year President Grover Cleveland appointed Shelby to the post of United States Marshal in Kansas City. As the new Marshal for the state of Missouri, Shelby shocked everyone by appointing the first black United States Deputy Marshal.

It may have seemed a contradiction to many for Shelby to remain friends with notorious characters and to stand up in open court in support of Frank James, one of the most notorious outlaws of the era. Jesse James had already been murdered, however, and Shelby may have felt that he still had a debt to pay for the loyalty of the James brothers to him during the war.

It had been at the bloody battle of Prairie Grove that Union Major James M. Hubbard overran Shelby's gun batteries with a hundred men. A confident Major Hubbard and his men had the Confederates surrounded and were holding Shelby at saber point.

"Surrender sir, or I fire," Hubbard had just said to Shelby. "You are surrounded."

At that moment Frank James, leading a band of Confederate guerillas including his brother Jesse and the infamous William Clark Quantrill, charged upon the scene turning the tables and startling Hubbard into a stunned silence.

Shelby smiled, then reached out and took the pistol out of Hubbard's hand saying, "You are mistaken, Major. It is you who are my prisoner."[8]

Known as a man of integrity, it should not be a surprise that Shelby would feel indebted to one who had been so loyal to him, even Frank James the outlaw.

CHAPTER TEN

Obituary

The Confederate colonies came to an end neither suddenly nor with a dramatic flourish. Many of the founders had seen in the plan the seeds of failure and were early deserters. The other colonists began to drift away soon after, either singly or in groups. The withdrawal of the French forces and Maximilian's abandonment of good judgment hastened the end which was already well on the way. As the year 1867 began, only the most resolute among the colonists remained in their settlements, and while the *Mexican Times* was printing optimistic encouragement, others were writing what they observed. An unknown correspondent for the New Orleans *Daily Picayune* posted the following dispatch from the Carlota colony on January 22:

> The colony of Carlotta is going—home...Only a year old, and already dead; going back to the states en masse...Today two families swarm the streets of Carlotta. The old country was not half as much abused going home as coming here. Universal suffrage at home is better than civil war and unlicensed robbers and hard work...
>
> One year ago General Price and Governor Harris selected this spot for a town...But the old general, the good old man, waited to see the bursting of the bubble and then went sadly home...Shelby waits at Cordoba for the completion of his Tuxpan enterprise...Only a few remain; the quiet men, not the visionaries and boasters, determining, somewhere here, to wait and see the end.[1]

Carlota, since it was the first and largest of the Confederate colonies, received the most attention in the contemporary press, but it is reasonable to assume that all the colonists shared the same spirit of disappointment. Familiar faces in Mexico, many of whom had served colonization in official capacities, were so scarce that the *Mexican Times* had taken to running almost an entire front page of advertising to take the space formerly occupied by news of prominent Southerners. Many of the colonists who had settled on "ex-appropriated" land found the former owners returning to claim their land as Liberal control increased in the outlying districts. Those who had not taken land but had come as merchants and businessmen fell under the provisions of a proclamation issued by Juarez in January 1867, forbidding them to operate. They had been friends of the Empire, and the country was speedily coming under the control of the Republic that recognized none of the acts or obligations of that Empire.

No doubt the main cause of the failure of the colonization program was the lack of a settled and certain form of government, one whose influence over the country could be depended upon. And perhaps the greatest mistake the Confederates made in establishing such an ambitious project was haste in choosing sides. Juarez had endorsed colonization before the Intervention and had repeated his support for such a program with the stipulation that it be carried out under the auspices of the Republic.[2] But the Confederates were men who had suffered four years of failure and were anxious for success. Maximilian's Empire seemed established in the summer of 1865, and he had shown them kindness in their exile and offered them homes. Few among them in their haste to leave the United States had presence of mind enough to realize the precarious position of a European-supported monarchy on the border of a self-righteous republic like the United States. In their frustration and defeat they had chosen and it was the wrong choice. They became identified with the usurper, the interventionist, the enemy of the Republic, and when the Liberals regained power the former Confederates were unwelcome.

> The great idea in the Mexican mind is to get rid of foreigners," wrote John Newman Edwards as the French were preparing to withdraw, "and whether it is the French this year, or the people of the United States next, it makes but little difference to them.[3]

The colonists might have expected the Mexicans to be against them, but few were prepared for the opposition of their own countrymen. As one newspaper correspondent wrote:

> The good men of the colony were put to shame by the reckless proceedings of those who came to obtain a living by their wits and their swindlings. The State prisons must have poured out the refuse here—the immoral, the robber, the libertine, the thief, all came here and fastened themselves with

> the citizens of America, North and South, and damned the colony and fleeced it.[4]

The most common misrepresentation encountered by the colonists who had come believing in the immigration literature, was that of the dishonest land speculator. These swindlers "rejoiced in fine outfits, flourishing and flashy signs that meant nothing; spent and played at faro…then left at night, in disguise—any way to evade the law and their victims—for the states," leaving the poor would-be settler with a worthless slip of paper entitling him to someone else's land.[5]…

When Carlota was finally abandoned the chroniclers of colonization declared it at an end. "The settlement of Cordoba is among the things of the past," John Newman Edwards wrote.[6] Then he too went home to Missouri to become a popular writer and newspaperman. The famous and near-famous Confederate officers and government officials returned home to the South to "share its fate" with Lee and Hampton. Governor Perkins went first to Paris before returning to Louisiana. Governor Harris left for Havana and then for home. Jubal Early forgot his oath against the United States and returned to his native Virginia to practice law. T.C. Hindman went home to his farm in Arkansas thinking he had escaped the uncertainties of life in Mexico, and was assassinated as he sat reading in his living room. Even Governor Henry W. Allen went home, a bit later than the others, and his remains were re-interred in a New Orleans cemetery.

The *Daily Picayune* published the Carlota colony's obituary and then closed the chapter on Confederate colonization in Mexico.

> The seeds of death were sown at the beginning…In the first rush an honest man could find not a foot of land, although fifty persons then held as many as a thousand acres. The speculators waited for high prices and they waited too long. They became generous when it was too late. They

> offered town lots and tracts of twenty acres gratis, when families and men by fifties took passage for New Orleans.
>
> The thatched roof bamboo houses on the Rio Seco, are deserted now—and the broad woodlands echo no more the axe stroke, and the familiar Anglo Saxon voice. The hoe and the plow, and the machete lie where the owner left them—in the cabin, on the ground where in the departure forgot them, or careless hands flung them—the relics of the rise and fall of colonization.[7]

The obituary was premature, but it predicted the final, total abandonment of Carlota. The settlers in other colonies were no doubt influenced by the experience of the Cordoba enterprise, but Maximilian had not yet quit the country and perhaps some saw in this, at least, some hope for the future. But even this hope vanished. Maximilian, the benefactor of Confederate colonization, died before a firing squad on June 19, 1867, with a prayer on his lips that his blood might be the last shed for Mexico. The colonist had reason to share his prayer.

General Escobedo, who commanded the siege of Queretaro, saw the emancipation of his country from foreign interference with the death of Maximilian and wrote the following to Governor Gomez of Nuevo Leon:

> I am now in favor of making clean work of the detested 'Gringos.' This country belongs to God and us, and just so long as one foreigner remains on our soil, our liberty is in jeopardy. By every means in our power we should make the country Mexican; and as all the property in the hands of foreigners was made by our misfortunes, we should take it, now that we have the power, and hunt them from

> the country. My motto now is, death to all estrangeros.[8]

By then most of the famous names had left Mexico. Jo Shelby, faithful to his promises, waited to see the last one of his friends on the homeward way; "then lingering behind as if protecting the retreat," went home to Kentucky and later moved to Missouri. The period of the great Confederate colonization was over. Thomas H. Nelson, United States Minister to Mexico wrote late in 1870:

> The large number of citizens of the Southern States of the Union who came to Mexico immediately after the Rebellion, have almost all retired to the United States…there is not at this moment in Mexico a single notability remaining out of the many Confederate refugees.[9]

Nelson was correct on two points: the "notabilities" had returned to the States, and his reference to the return of "almost all" the "citizens of the Southern States" correctly implied that some had remained in their Mexican colonies. One who remained to serve up the news for Southern readers was George W. Clarke. He established the *Two Republics*, another English language newspaper to replace the abandoned *Mexican Times*. A former newspaper editor in Van Buren, Arkansas, and later a major in the Confederate army, Clarke had come to Mexico among the first expatriate arrivals, and served under Maury as a colonization agent in Durango.[10] His durability and adaptability to his new environment is attested to by the fact that he remained as publisher of the *Two Republics*, until sometime after 1878.[11] Beginning the publication as he did with the restoration of Juarez, Clarke adjusted his editorial policy to the philosophy of the Mexican Republic, hence its title.

The Confederate bias of Clarke's newspaper, however, is unmistakable. The news he printed and the articles he selected from foreign papers for reprint dealt largely with items pertaining to the South and appealing to the Southern expatriates in Mexico. He frequently mentioned the names and activities of former Confederates, thus providing future historians with one of the few evidences of a sizeable Confederate population remaining in Mexico after the evacuation of Carlota and the decline of immigration after 1867. Although once again it is impossible to positively identify every American as a former Confederate, it appears that most of those Clarke wrote about were of the same origin as he.[12] That there was an extensively disseminated Confederate population in Mexico after the reinstatement of the Republic is borne out by the list of agents for The *Two Republics* who operated in widely separated areas of Mexico: Dr. Sutton at Orizaba, Alfred Lewis at Durango, Major J. B. Russell at Monterey, and a Dr. Knight at Saltillo.[13]

The *Two Republics* in the first few months after its beginning carried frequent news of Southerners in Mexico City. In August, 1867, an item appeared concerning a banquet recently given in the capital which was attended by General Slaughter, Colonel Cripps, Colonel Green, Colonel Blake, Captain Proudfoot, Major McMurty, Shelby's former business partner, Captain Richard Taylor, and Major Trailkill and Captain Wood, who had come to Mexico with Shelby. The article does not disclose the occasion being celebrated, but as reported in the article, it amounted to quite a gathering of the former Confederate officer corps.[14]

A short time later another article appeared in the pages of The *Two Republics* that spoke of resident Southerners. The item provided a list of "American citizens" who had signed a petition for the appointment of one of their number to the vacant post of United States Minister to Mexico. The American they supported for the appointment was a lawyer, Frederic Hall, who later wrote a history of the Intervention and Maximilian's Empire. Among those signing the petition were: George M. Green of California, J.M. Schnierle of South Carolina, Thomas Griffin of New Orleans, E.R.

Wells of Texas, Harry Levy of Louisiana, and W. E. Jowell of Virginia.[15]

Governor T.C. Reynolds, of Missouri, had remained in Mexico City as John Newman Edwards had predicted he would,[16] and The *Two Republics* devoted a considerable amount of space to his activities. Clarke mentioned the Governor in an article on President Johnson's amnesty proclamation, pointing out that Reynolds was the only Confederate who had not yet been pardoned.[17] A later article announced Reynolds' appointment as general superintendent of the Mexican Railway Company, which would operate the line from Vera Cruz to Mexico City.[18]

Another item concerning Reynolds which also pointed out clearly that Juarez was not completely opposed to the presence of the Confederates in his country was a reprint of a letter that had originally appeared in the St. Louis *Republican*. Reynolds had written the letter telling about a recent meeting with President Juarez himself. In the letter Reynolds praised the Juarez government for its friendly treatment and recalled that Juarez had given him "a most courteous and cordial reception by inviting me to consider myself in Mexico 'not as an exile; but in my own land and home.'"[19]

Judging from the articles cited it is obvious that not all the Confederate exiles had made their way back to the United States after the fall of Maximilian's Empire, nor was the idea of emigration dead in the South. With the disappearance of The *Mexican Times* interested parties began to address their requests for information to Clarke's newspaper instead. These letters came in such numbers and became such a nuisance that he notified his readers that requests for information would be honored in the future only if the writer enclosed a "greenback."[20]

The letters coming into Clarke's office did not represent idle interest on the part of their writers. Notices were printed as late as December of Texans departing from Mexico City for the colony Shelby had founded at Tuxpan.[21] One of these later immigrants left his memoirs describing the venture under the Juarez government.

Gideon Lincecum moved to Tuxpan from Longpoint, Texas in the summer of 1868, and lived there for several months before returning to Texas. Shortly after his arrival he wrote of the American population in Tuxpan:

> There are, I think, forty-two American families living in the Tuxpan Valley, four on the south side of the river below, and the balance above Tuxpan. Several of those who settled as high up as Tumbadero, preferring the lower portion of the valley are selling their places and moving down here…
>
> Besides the lazy malcontents, there are a number of substantial families in the valley who are doing very well. The signs of prosperity are manifest in the fact that they are clearing and setting all the land they can in sugar cane and coffee. They are men and will do well in time to come.[22]

How prosperous the Tuxpan colonists had become is not told, but if they remained long enough and claimed title to their land long enough, they might have become wealthy in the next fifty years, when the rich oil pools beneath the agricultural soil were tapped.

Clarke also received a request for information from a Southern African-American who was interested in establishing a colony for black families in the *Tierra Caliente*, which had been avoided by the earlier colonists. The *Two Republics* printed the following notice of the letter:

> Francis Durant wishes to know if 'liberal arrangements' can be made with the Government of Mexico, to colonize a number of 'brethren' (colored) in the tierra caliente where he feels assured their 'constitutions, temperature and habits'

> are adapted and they can enjoy a 'life of ease and happiness.'[23]

To another inquiry from a printer who wanted to know if he could make a living in Mexico, Clarke answered:

> …it depends on how he wants to live. If he can be content with frijoles and tortillas with a little variation of red pepper occasionally he might get along here. Or in other words wages are low, as the wages of all mechanics are; but expenses are lower than in any other capital[s] that we know of.[24]

Still another piece of evidence that the old interest in colonization was still alive is a *Two Republics* article concerning a recent venture by Hamilton P. Bee, late of the Confederate army, and Dr. D. McKnight. The two Texans leased a hacienda near Pareas in the State of Coahuila where their activities would include "agrarial experiments, paper manufacturing, vineyards, and wine making and brandy distilling."[25]

In spite of the favor shown by the Juarez government to George W. Clarke, Governor Reynolds and others, the colony at Tuxpan had frustrating problems. Major John H. Brown, one of the many Texans who were still living in the Tuxpan district three and a half years after the first colonists arrived there, wrote to the United States Minister to Mexico, Thomas H. Nelson, in September, 1869, describing the conditions in his neighborhood. He said that most of the settlers were:

> poor, though a few possessed means sufficient to open and cultivate respectable farms. They are generally steady, industrious and law abiding; and have lived as resident American citizens in entire peace with the natives, without a single disturbance

> between them, or a personal outrage to complain of.[26]

However, Major Brown's purpose in writing was to complain about the treatment the colonists were experiencing at the hands of the customs officials. They had suffered "exactions and delays" and were especially being heavily taxed by "duties on tools, implements and machinery" brought by the settlers "to aid in developing a section in which neither a wagon, plough nor hoe had ever been sued by the natives." Such unfair treatment on the part of these government officials was working an undue hardship on the colonists, Major Brown reported. He then proceeded to give some pertinent examples.

The enterprising settlers of the colony had decided to erect a sugar mill and proceeded to import the parts by installment. On the arrival of the first small installment the colonists were required to pay heavy duties on the whole machine, the last parts of which would not arrive for another six months. Though Major Brown does not say so, the settlers were probably wondering if, after paying such exorbitant duties they would be able to afford to put the mill into operation.

Another example given was that of a widow whose profession was teaching music. In order to practice her profession she imported a piano that she was forced to sell to pay the $100 duty demanded by the customs officials. Major Brown himself had been charged a fee of $12.80 on a box of seeds, intended for free distribution among the other colonists, which had only cost $10 in Galveston. The Major lodged such a strong protest that the seeds were finally allowed into the country duty free, but a frustrating delay followed before the seeds were delivered. Major Brown's letter suggested that this harassment "occurred, as if designed to deter foreigners from locating in that fertile and wilderness country."

Having stated his grievances and recorded the usual pleasantries, Major Brown closed his letter with the following postscript:

> With some reluctance I venture to allude to the impression to some extent existing in our native land, that this settlement is composed of disaffected ex-southerners. I cannot speak for the entire little community, because I am not intimate with all, but if there is a single one who wishes to denationalize or un-Americanize himself I do not know it—nor do I believe there is one such. On the contrary, though a large majority hail from the South, they have come here chiefly to recuperate from the ravages of the war, and they earnestly pray for the happiness, prosperity and glory of their native country in all its length and breadth. Moreover [,] a fair portion are from the North, including several married ladies, (like my own wife) who are not mentioned.
>
> We have as earnestly desired emigrants from the North as from the South, regarding them as of the same blood, language, religion and free institutions, and realizing that, though lately engaged in dreadful and lamentable strife, they are nevertheless the only people under the sun combining all the great elements of affiliation with us as citizens of a free, representative Republic.[27]

Could the sentiment expressed in this postscript mean that while the Reconstruction further embittered the ex-Confederates of the South; the ex-Confederates of the Mexican colonies had healed their wounds in three years of wilderness pioneering in a foreign land? Perhaps the Major was being tactful as there was a suggestion accompanying his letter, signed by fifty-four Americans, that an American consul was needed in the colony to assist the settlers in obtaining better treatment from the Mexican authorities. The records do not say whether their request was granted, but is probably was not.

After this letter the records become silent concerning the Confederate colonists of Mexico. The *Two Republics* continued publication until near the end of the nineteenth century, still giving preference to items of Southern interest, at least until George W. Clarke passed on the editorship to another. The last mention of any organized Confederate colonization is made in an article entitled "The Bogus Bureau of Mexican Emigration":

> There is no foundation for the rumor in St. Louis that Mr. Marischal, the Mexican minister, has made a contract with Generals Shelby and Bacon Montgomery for emigration to Mexico. Mr. Marischal says he has no instructions to enter into such agreements, and there is no official agency for immigration to Mexico established in the United States of any other country that we can hear of.[28]

Shelby was apparently popular with Clarke, who published a substantial item containing the general's views on the current "Mexican problem."[29] Clarke also took occasion in the last months before his name vanished from the editorial page to reprint favorable items about former Confederates in the United States and to take a swipe at General Sheridan.[30] Sometime after 1878 The *Two Republics* came under the direction of J. Mastella Clarke and Walter M. O'Dwyer, and the Southern sympathy disappears.

Obviously all the Confederate exiles did not abandon Mexico with the failure of Carlota and the other colonies. They had hoped to establish the Old South in Mexico, where the happy days they remembered from before the war could be lived again. Their hope vanished and hundreds, perhaps thousands, came and returned within a few months. We cannot assuredly say how long George Clark's officer friends and Major Brown's colonists remained after their newspaper and their letters cease, but men do grow to love a country and they put down roots when they rest upon its soil long enough. Only imagination can take the story from there, but some undoubtedly lived out their lives of exile in their adopted country as they had lived out their great adventure: unreconstructed, their spirit unbroken.

CHAPTER ELEVEN

Afterward

Southern interest in Latin American was not aroused as an outcome of the Civil War; indeed, it reached deeply into the era of Manifest Destiny. Men such as John Mason, William L. Yancey, and Matthew F. Maury had brought to the attention of Southern and national leaders the suitability of Mexico, as well as Central America, Brazil and Cuba, for the expansion of slavery. Others like Robert J. Walker looked to Mexico as a safety valve to drain off surplus slaves, eventually creating a neutral and emancipated race in Latin America as African-Americans intermarried with the native populations. The filibustering expeditions of William Walker and others prior to the Civil War further stimulated interest in Latin America as a logical area for expansion. In discussions of national expansion the term Manifest Destiny was applied as often to southward expansion as to the westward movement, as evidenced by the space devoted to the region by leading newspapers and periodicals during the fifteen or so years preceding the Civil War.

Matthew Maury, who became the father of Confederate colonization in Mexico, had looked southward to Latin America as the most logical direction for United States expansion as early as 1850. To him the Gulf of Mexico was the Mediterranean of the Western Hemisphere and the destiny of the United States was to control the Gulf and the Caribbean by eventually annexing the Latin American lands touched by those waters. To establish a claim for annexation, Maury reasoned that Americans, especially Southerners, should establish settlements in these areas. Such settlers could, after a time, agitate for American control in much the same way Texas gained entrance into the Union.

With the Mexican-American War the South had established a reputation for coveting Mexican territory, which sustained itself through the difficulties of the 1850s. It was this expansionist reputation, which made Juarez suspicious of Confederate motives when Pickett was sent to Mexico to obtain diplomatic recognition for the Confederacy in 1861. Juarez believed that the most important consideration in earlier American expansion into Mexican territory had been slavery. Holding the South responsible for this he remained friendly to the United States, believing that with the issue of slavery removed from the discussion there was no danger of further territorial disputes from Lincoln's government. Pickett had added to Mexican suspicions by indiscreetly suggesting that his government exploit Mexican friendliness toward the United States as a pretext to invade and annex part of Mexico to the Confederacy. The Confederate Government, however, had no interest in acquiring any of Mexico's territory, and unhappy with the trouble being stirred up by their representative, recalled Pickett from his post.

As the war continued it became apparent to many leaders throughout the South that slavery would have to be abolished. Therefore when the Confederate generals turned their eyes toward

Mexico in 1865, it was for refuge, not for territorial acquisition. When Maximilian temporarily replaced the Juarez government he, as well as the French who supported him, were friendly to the South and extended to the Confederate exiles a liberal proposal for colonization. Under the direction of Matthew Maury, who had modified his earlier colonization plan, public and private lands were surveyed and set aside for settlement by Southerners who came to Mexico by the hundreds in the next eighteen months. Some were embittered against the United States; others had lost everything in the war and were eager for a new start. Still others, like General Alexander Terrell, hoped to offer a large portion of Mexican territory occupied by Southerners to the United States; though that view seems not to have been popularly held among the majority of the exiles.

Maury became the great publicist of Confederate colonization, and was assisted in disseminating immigration information by Henry Watkins Allen through his English language newspaper, The *Mexican Times*. Although Allen tempered his claims of a Southern Utopia in Mexico, Maury exaggerated wildly and painted such a picture of heaven on earth that multitudes of would-be settlers swarmed into Mexico naively expecting to find large tracts of free land, which readily yielded its bounty, as had the Garden of Eden. The rhetoric Maury used in calling settlers to Mexico is reminiscent of that used in the great surge westward before the Civil War, and it would be heard again as the westward movement resumed once the war came to an end. Maury promised the immigrant to Mexico a twelve-month growing season and a perpetual harvest springing forth from land so fertile that only a token effort at cultivation was necessary. His descriptions of the land offered for colonization by Maximilian outdid the hyperbole used in earlier promotions calling immigrants to California. In neither case were those expectations fully realized; the lands were excellent, but they weren't the Garden of Eden—or even Oleana.

The invitation to settlers bound for Mexico was aimed primarily at Southern farmers, and an implied promise was held out of rebuilding the old aristocratic way of life. A utopian state of mind had developed in the Old South of a perfect agrarian society where everything and everyone had a specific place. The old Southerners continued to depict their new lives as duplicates of the old when they moved west. The South had cast its society in the mold of the English gentry and claimed direct Anglo-Saxon descent, as did no other section of the country. Therefore the Southerner saw himself as the natural leader of Manifest Destiny, carrying the civilizing influence of Anglo-Saxon culture into new lands.

The Confederate colonization experience in Mexico drew upon this background in the official rhetoric of the movement. In this area it was Henry Watkins Allen who expressed the philosophy of the leaders of colonization. Mexico was pictured as a backward land that had been passed over by civilization and greatness owing to the absence of the Anglo-Saxon influence, which had already proven itself in the American experience of growth and expansion. As a further proof of the theory, those lands so recently belonging to Mexico, especially Texas and California had flourished and prospered in the few short years of American administration. Allen foresaw the cure for all of Mexico's ills in the immigration of large numbers of Anglo-Saxon Southerners who would bring modern agricultural knowledge and methods to regenerate the economy. Allen's editorials in The *Mexican Times* proclaimed the high-minded purposes of Confederate colonization as the greatest boon to Nineteenth Century Mexico, save the establishment of Maximilian's Empire. Not only would the colonists introduce an agricultural revolution into Mexico, but they would create a middle class in a land where none existed, thereby raising the standard of living for all Mexicans. Matthew Maury, through his propaganda, gave the discontented Confederates of the ruined South a reason to immigrate to Mexico; Allen gave them a cause.

The myth of a Southern utopia died hard, and the Confederates brought that idea with them to Mexico. Now reconciled to the notion that slavery was forever abolished, they reflected upon the great agricultural economy of pre-war days and determined that the freed slaves could be employed in Mexico to reestablish the supremacy of Southern agriculture. Laboring there as free men under their former masters they would be regarded as apprentices who would, after a given period, take their place in the new Southern social and economic system as trained farmers capable of contributing gainfully to that society. Thus the orderliness of the antebellum South would be preserved and the United States would have to concede defeat in its effort to destroy Southern civilization. The radical Republicans could rave on with their Reconstruction, but the best of the old order would reestablish itself in Mexico and prove its endurance and adaptability.

That preoccupation with orderliness that had characterized the myth of the old Southern society became an obsession with the Confederate exiles in Mexico. Their haste to flee the destruction of the South and to reestablish their way of life in Mexico brought them into conflict with a set of realities they did not understand. Order in Mexico had been a goal of its many revolutions but never an accomplished fact. To the Southerners Juarez represented anarchy, chaos and disorder. John Newman Edwards persisted in referring to the Liberal troops as guerrillas and bandits, expressing the general Confederate attitude toward the Mexican Republic. Maximilian, on the other hand, symbolized that feudal order to which they felt they were accustomed. Mexico through a series of pointless bloodbaths had proven itself unready for a republican form of government, and was in need of a period of tutelage that could be provided by Maximilian's regime. The Confederates, therefore, supported the Empire.

It appears that they expected an unusual arrangement concerning their role in Mexican affairs. Preferring law and order to the chaos, which they felt was sure to come with a reinstatement of the Republic, they seem to have expected a certain amount of autonomy in their colonies. Having no faith at all in Juarez, the Southerners perhaps trusted too much in the ability of Maximilian to insure the order they sought. Opposed by the United States, ill-advised by his counselors, and deserted by the French, Maximilian could not control even his own fate, much less the Empire which was expected to bring the benefits of civilization to Mexico. Unprotected by either side and not strong enough to defend their own settlements, the colonists found themselves caught up in the anarchy and disorder they had hoped to see ended as a result of Maximilian's rule.

Although some resolute colonists remained in Mexico and a few more left the United States to join them, the Confederate colonization movement as Maury and others had envisioned it, came to an end in the summer of 1867. The project had been established for many reasons, just so its end was the result of several factors. The success of the program, regardless of its own inherent weaknesses, depended upon the ability of Maximilian's regime to control the country. The Confederate military units that began arriving in Mexico in the early summer of 1865 had offered the emperor his best opportunity to firmly establish his government. If it were within the realm of the historian to speculate, it would be interesting to trace the course history might have taken had Maximilian accepted General Shelby's plan to recruit 50,000 Americans to serve as his main military force. But Maximilian was not a decisive man and was perhaps the last to realize the significance to his government of a loyal force of Confederate veterans. He was genuinely concerned with the welfare of Mexico and honestly believed that his government could give the Mexican people the peace, justice and prosperity they desired. In short, he was a young idealist caught in the midst of a set of circumstances spinning ever faster out of control. He considered himself a Mexican and his cause a Mexican affair; hence he refused to accept Confederate soldiers into his army.

As for colonization itself, Maximilian's failure to follow the project through and to provide adequate land for settlement made the task of the Office of Land Colonization almost impossible. Maury's exaggerated publicity brought countless immigrants to

Mexico, who once there found not only a shortage of free land but unscrupulous speculators demanding outrageous prices for private land. Having come to Mexico at such great personal sacrifice they no doubt experienced a sense of betrayal when faced with the humbug of the Emigration Office's overstatements. Having few other options they returned to the Untied States with discouraging reports that checked further interest in emigration.

Mexico did not prove to be the El Dorado Maury had pictured it to be, nor were conditions favorable for reestablishing the antebellum life. With the withdrawal of the French and the decline of Maximilian's control of the country, harassment of the colonists by Juaristas and bandits destroyed the last remaining hope of success.

Another factor which must be noted in the failure of the Empire, and which in turn contributed to the passing of colonization, is the role of the United States in aiding Juarez and forcing the French withdrawal. The attitude of the United States Government had changed considerably since Blair's mission to Richmond in January of 1865. The proposals made by Blair provide another tempting area of speculation.

Blair had made it clear, perhaps from his own understanding of the proposal, that Lincoln would not oppose a Confederate-led joint expedition to expel the French from Mexico. And in that regard President Jefferson Davis seemed to understand that, once his Confederate States were returned to the Union, former Confederates might, by joining in such an expedition, play a major role in extending the boundaries of the United States to the "Isthmus of Darien." A provisional government could then be established which would "model the Mexican states so as to adapt them to the Union." Though the peace negotiations, which resulted from Blair's mission failed, such an expedition seems to have still been under consideration and Shelby led his troops into Mexico under that assumption.

Whether Shelby and the other Southern luminaries had personal dreams of empire can be argued, but if such imaginings motivated them in their flight into Mexico it may well be that by refusing to join Juarez, Shelby's men denied the future annexation of Mexico to the United States. Perhaps the chance happenings of history force upon men the wisdom they refuse to exercise. Such a move could have been tragic had another war with Mexico ensued. By attaching themselves to Maximilian's Empire as colonists rather than soldiers, the Confederates made it possible for Seward to take a less dangerous and less expensive path in forcing French withdrawal by materially and financially supporting Juarez. In this role of freedom's guardian, the United States finally abandoned the southward drift of Manifest Destiny.

As for the former Confederates, their experience with colonization in Mexico at least made them more appreciative of the Republic they had left in the United States, for they returned home convinced that the kind of life they wanted could be found nowhere else. The South was still their home, and Yankee occupation troops and radical Republicans seemed small obstacles compared to the chaos and bandits and strange diseases of Mexico. A general discontent with things as they were in the South following the war had impelled the colonists to seek their shattered hopes in Mexico, and if they did not come to realize that such a utopian dream might never have existed, at least they learned that it could not be exported. For the majority the colonization experience had been as painful as the war, and the Reconstruction they had tried to escape seemed a welcome relief.

When the colonists returned home, they joined the majority of Southerners who had remained in the United States in declaring the causes of the war resolved. They seemed as willing to forget their experiences in Mexico, as they were to resume life in the South they had fled. Benjamin H. Hill, though not a colonist, summed up the general Southern opinion when he wrote in 1876:

> Let the people of each section admit the great truth which will demonstrate the manhood and worth of both: That we differed honestly, that we

fought bravely, and that our differences are settled, in good faith, on the basis of the Constitution as it is. And now, with our people all free, our States all sovereign, let us all unite our strength to cause our "Father's House" to become the light of the civilized world.[1]

A COLLECTION OF BIOGRAPHICAL SKETCHES OF MAJOR PARTICIPANTS IN THE STORY

MEXICAN PARTICIPANTS

Santos Benavides

November 1, 1823 – November 9, 1891

Santos Benavides was born in Laredo, Texas, the town founded by his great-great grandfather, Thomas Sanchez de la Barrera y Garza. As a young rancher and political leader in the Laredo border area Benavides helped bring a traditionally remote area into the mainstream of Texas politics. Eventually, in 1856, he was elected mayor of Laredo and chief justice of Webb County in 1859. He established a firm reputation as a leader when he was instrumental in bringing to an end the lawlessness of the Rio Grande region and in fighting against the Lipan Apaches on both sides of the border.

When Texas seceded from the Union, Santos and his two brothers, Refugio and Cristobal, joined the Confederate Army (over 2500 Mexicans fought for the Confederacy) and formed the Thirty-third Texas Cavalry. Benavides was soon commissioned a captain and won accolades for his fighting skills. On May 22, 1861 at the Battle of Corrizo, he drove Juan Cortina's renegade army back into Mexico killing several and executing several others. After quelling many renegade revolts and uprisings up and down the Rio Grande, and driving back Indians and Juarista raiders, he was promoted to major, making him the highest ranking officer of Mexican descent in the Confederate Army.

Benavides' cavalry company crossed the Rio Grande on September 2, 1863, and cornered the notorious bandit (and Unionist) Octavio Zapata near Mier, Mexico. Zapata was associated with Colonel Edmund J. Davis, a Union officer operating in the Matamoros-Brownsville area. Zapata often flew a U.S. flag while on his raiding expeditions and referred to his group as the "First Regiment of Union Troops." Benavides' Confederates engaged Zapata's troops and, after a brief gun battle, left over ten bandits dead, including Zapata himself. It is said that it was Benavides who personally killed Zapata.

In a later significant engagement fought on March 19, 1864, Benavides won a nightlong battle against Edmond Davis's First Union Texas Cavalry just east of Laredo, along the banks of Zacate Creek.

Before the war ended Benavides rose to the rank of colonel. Perhaps his greatest contribution to the Confederate cause came during the Union blockade of Brownsville in 1864-65 when he was involved in protecting the movement of Texas cotton along the Rio Grande to Matamoros. Colonel Benavides is the only Confederate officer to have fought against Union forces, Mexican Army troops, and Apache Indian tribes.

Benavides finished out his life as an influential politician and alderman of Laredo in recognition of his political and military achievements. Benavides was selected as an envoy to the United States by Porfirio Diaz, and by the state of Texas as a delegate to the World Cotton Exposition. Colonel Benavides died in his home in Laredo on November 12, 1891.

Juan Nepomuceno Cortina

May 16,1824 – October 30, 1894

Juan Cortina, sometimes known as the "Robin Hood of the Rio Grande," was born the son of aristocratic parents in Camargo, Tamaulipas, and heir to their large fortune and land grant in the lower Rio Grande Valley near Brownsville, Texas. Cortina served as a cavalryman in the Mexican Army during the Mexican-American War participating in the battles of Resaca de la Palma and Palo Alto. After the war he returned to the banks of the Rio Grande and to the family ranching business, but soon ran into trouble with local authorities when he was accused of cattle rustling. Only the influence of his powerful family saved him from prosecution.

Cortina's anger over the loss of the war to the *Gringos* affected him for the rest of his life and he was never able to come to terms with what he saw as the arrogant attitude of the Anglos living in the Brownsville-Matamoros area along the lower Rio Grande. A significant turning point in his life came on July 13, 1859 while on a business trip to Brownsville. Cortina witnessed Sheriff Robert Shears administering a pistol-whipping to one of his former vaqueros in the process of making a supposed lawful arrest. When Cortina intervened, a gunfight erupted which resulted in

Sheriff Shears receiving a gunshot to the arm. In the confusion following the confrontation Cortina and the vaquero managed to escape, but the incident was the beginning of a new career for Cortina.

The shootout in Brownsville ignited the first "Cortina War" which began in earnest on September 29, 1859 when Juan rode back into Brownsville with about 50 men and seized control of the town. Several casualties occurred as Cortina's men rode through the streets shouting, "Death to the Gringos." After holding the town for a few days, Cortina's guerillas withdrew to the family ranch at Santa Rita in Cameron County where they were attacked by a group of townspeople from Brownsville and local volunteers. After a brief battle the vigilantes were driven back.

Cortina's outlaw reputation grew and in short order the Texas Rangers took up the responsibility for hunting down Cortina and his band. Fierce fighting that took place over the next few weeks caused significant loses in civilian property on both sides of the Rio Grande and threw the entire area into chaos. A company of Texas Rangers led by Major John S. Ford and Major Sam Heintzelman handed Cortina's army a severe defeat on December 27, 1859 at the Battle of the Rio Grande and pushed him into Mexico.

Still defiant Cortina and his followers returned to the lower Rio Grande where they captured the steamboat *Ranchero*. The Rangers quickly retaliated and eventually forced Cortina into the wilderness of the Burgos Mountains where he remained until Texas seceded from the Union in 1861. With Texas caught up in the excitement of the Civil War he began the "Second Cortina War" in an attempt to seize the town of Corrizo in Zapata County.

When Cortina invaded Zapata County in May 1861, the Confederates decided to take action and dispatched Captain Santos Benavides and the Thirty-third Texas Cavalry to find and eliminate Cortina and his small army of guerilla outlaws. Benivedes located Cortina on May 22, 1861, and killed seven of the outlaws in the ensuing battle. Eleven others were captured and executed, either by hanging or firing squad.

Following this defeat Cortina allied himself with Juarez and, by default, with the Union cause. However, his ability to conduct raids upon Confederate supply trains and towns along the border were severely diminished. The French intervention in Mexico diverted Cortina's attention from his American pursuers to the French invaders. He went on to serve with the Juarista forces at the Battle of Puebla and continued to see action against the French in central Mexico. In 1863 Cortina proclaimed himself governor of Tamaulipas and was promoted by Juarez to general in the Mexican Army of the North. He later participated in the siege of Queretaro and was present when Maximilian was executed by firing squad along with his two loyal generals, Miguel Miramon and Tomas Mejia, on June 19, 1867.

Cortina returned to his family ranch near Brownsville in the 1870s and attempted to clear his name. Porfirio Diaz was dictator president of Mexico at the time and under tremendous pressure from the United States to be more circumspect in his political dealings. Partly in a gesture to the Americans Diaz had Cortina arrested and imprisoned for four years. Following his release from prison Cortina, defiant to the end, died in Alzcapozalco on October 30, 1894.

Jose de la Cruz Porfirio Diaz Mori

September 15, 1830 – July 2, 1915

Porfirio Diaz was born in the city of Oaxaca, not far from the birthplace of Benito Juarez. In 1843, young Diaz was sent to a strict Catholic school, the Seminario Conciliar. However he found the priesthood too restrictive of the life style of an adventurous sprit such as himself and he left in 1846, later joining the military. His enlistment was too late for him to participate in Mexico's defeat at the hands of the United States in 1848.

Diaz was practicing law in Oaxaca when French troops arrived in Mexico and prompted him to again join the army in 1862 to oppose the invaders. He led troops in the victory over the French Army at Puebla on May 5, 1862; memorialized to this day as *Cinco de Mayo*. The French defeat at Puebla was the first time the French Army had lost a battle since June 18, 1815, when Napoleon was defeated at Waterloo, Belgium by the Duke of Wellington at the head of a coalition of European troops.

Diaz was captured by French troops in 1863, but soon escaped to later become Division General of the Central Army of Mexico. Captured again by the Imperial forces he was offered amnesty and a chance to join their cause; an opportunity he declined. Soon after he escaped once again and went on to lead Mexican troops in the battles of Tehuizingo, Piaxla, Tulcingo, and Comitlipa. He formally declared his allegiance to Benito Juarez in 1866 and went on to win more victories at Nichixlan and Miahuatlan. During the Mexican-French War Diaz was twice seriously wounded, imprisoned on three different occasions, had three narrow escapes from the enemy, and took part in over ten engagements against French troops.

When Diaz lost election to the presidency against Benito Juarez in November 1871, he claimed that there had been widespread corruption in the election process and began a revolution against the Juarez government. When Juarez died on July 9, 1872, Diaz accepted the presidency and helped to reunite the country by offering amnesty to troops loyal to Juarez. A cunning politician he maintained power by whatever means required, earning for himself a menacing reputation between both friend and foe. Perhaps as a political ploy he stated in 1908 that Mexico was ready for democracy, but he never stepped down until forced to do so as a result of the Mexican Revolution (1910-1911).

Diaz finally resigned from office on May 25, 1911, after the collapse of the government and made one last escape, this time fleeing to France. He died there an exile on July 15, 1915. Diaz lies buried in the Cimetiere du Montparnasse in Paris.

Benito Juarez

March 21, 1806 – July 18, 1872

Benito Juarez, a Zapotec Indian, was born into the poverty so typical across Mexico. He still only spoke the Zapotec language when he began school. With his heritage, his language, and poverty dominating his young life he rose to be a reformer and eventually President of Mexico.

As Juarez became interested in politics he gave his allegiance to reforming his nation culturally, politically and economically. As he became more politically astute he became an admirer of the government of the United States and by the time he became president he also had great respect for an American who would become president of his own country at the same time Juarez was president of his: Abraham Lincoln.

Mexico had declared its independence from Spain in 1821, but the attempts at self-government had given way to bloody uprisings, civil wars, and political intrigue. Antonio Lopez de Santa Ana had been involved in the original movement to achieve independence for Mexico from Spain, but he seems to have had a character flaw of Greek drama proportions and took advantage of the unstable environment of the country in his rise to the presidency of Mexico at least an eventual six times. During each rise to power he acted more like a dictator than an elected leader. Juarez became more devoted to the cause of meaningful reform of Mexico and became one of Santa Ana's political enemies.

Juarez served as governor of the state of Oaxaca from 1847 until 1852. His growing opposition to the corrupt national leadership of Santa Ana caused him to flee Mexico in 1853 to escape the Mexican dictator's wrath. He worked in a New Orleans cigar factory while in exile. While an expatriate he helped to draft the Plan of Ayutla calling for Santa Ana's resignation and a convention to create a new constitution.

Santa Ana resigned in 1855 and Juarez returned to Mexico. A provisional government under General Juan Alvarez brought in the period known at La Reforma. The Reform rescinded the powers of the Catholic Church and eventually, when Juarez became president he confiscated not only Church lands but also much land from the landed aristocracy. Under the new constitution of 1857 Juarez became President of the Supreme Court of Justice under new president Ignacio Comofort, followed by general political instability.

Though the Conservatives proclaimed General Felix Zuloaga president, under the 1857 Constitution Juarez, as President of the Supreme Court of Justice, became the interim president and was acknowledged president in his own right on January 15, 1858.

He now became the leader of the Liberal side in the civil war known as El Guerro de Reforma (Reform War). General Zuloaga's troops held the capital of Mexico City and forced Juarez and his followers to flee eventually to Vera Cruz.

With the unofficial support of the United States Juarez' forces captured Mexico City on January 1, 1861, and he was chosen in what passed for a general election as the official President of Mexico. Among other difficulties the nation suffered extreme economic problems. In an austerity plan Mexico suspended payment of foreign loans, which led to the Tripartite Treaty between Spain, France and Great Britain. The stated plan was for the three European nations to control the Mexican customs houses until the debts owed them had been paid. This was pressed by Louis Napoleon III of France in an effort to involve Spain and Great Britain in a general intervention in Mexico's affairs.

Spain and Britain quickly saw through Napoleon's plans and withdrew their troops. Napoleon needed a foreign success to reassure his countrymen of his regime's power and sent even more French troops to attempt a take over of Mexico on their own. To add legitimacy to his plan he convinced the younger brother of the Emperor of Austria, Maximilian von Hapsburg, to accept the title of Emperor of the Empire of Mexico.

Again Juarez was forced to remove his government to the northern provinces of Mexico and even spent time himself during this period in the United States. When the American Civil War, which had distracted Juarez in his attempts to force Maximilian from the throne of Mexico, ended the United States began to put pressure on the French to leave Mexico. In Europe the rise of Prussia threatened a war with France and further influenced Napoleon to retreat from Mexico.

Juarez gave Maximilian an opportunity to leave Mexico with the French, but he was persuaded by his loyal Mexican generals to stay and continue to extend his tenuous control of Mexico. Maximilian and his dwindling army of loyalists were defeated at the Battle of Queretaro in June 1867. Juarez rejected international pleas for amnesty for Maximilian and his safe return to Austria, and the one-time Emperor of Mexico was executed by firing squad on June 19.

Juarez was re-elected in controversial ballots in 1867 and 1871 in which he was forced to follow a typical procedure for the time; he used the power of the office to guarantee his victories. Revolts led by Porfirio Diaz and others continued to tax his energies. He died at his desk in the national palace as he sat reading a newspaper on July 18, 1872. His Foreign Minister, Sebastian Lerdo de Tejada, succeeded him.

Santiago Vidaurri

July 25, 1809 – July 8, 1867

Santiago Vidaurri was born July 25, 1809, the son of Pedro Vidaurri, a powerful caudillo, in the northeastern Mexican town of Lampazos, Nuevo Leon. Santiago went to work as a clerk for the Nuevo Leon police and eventually became the chief assistant to Governor Joaquin Garcia in 1837. He also served as secretary for Governor Manuel Maria de Llano and, as a very young bureaucrat, went to Texas to spy on Mirabeau B. Lamar's Texan Santa Fe Expedition. From 1840 to 1850 he worked for the conservatives in the Mexican government. After the loss of the Mexican American War in 1848, Vidaurri denounced the regime of Antonio Lopez de Santa Anna and, with his rebel army, captured Monterey on May 23, 1855, and was installed as the military governor of Nuevo Leon. In an attempt to establish a new liberated and independent nation he called "The Republic of Sierra Madre" Vidaurri soon moved into the states of Coahuila and Tamaulipas, gaining control over much of northern Mexico.

When the American Civil War began Vidaurri actively supported the Confederacy and cooperated by using his troops to assist in punishing Indians and renegades who raided Confederate supply trains. He also assisted in the apprehension of fugitive slaves in Mexico. These activities did not endear him to his liberal enemies, nor, in some cases, to his conservative compatriots who did not trust the Confederates any more than they did other foreign invaders. Nationalists like Benito Juarez denounced Vidaurri, and even several of his own officers, who distrusted his associations with the Gringos, defected to the National movement.

In the volatile arena of Mexican politics of the day Vidaurri had little choice but to support Maximilian and the Imperialists, who in turn winked at efforts to assist the Confederacy to the north. Vidaurri continued to be instrumental in protecting the underground Confederate supply lines moving through ports on the Mexican coast to avoid the Union blockade of Confederate ports. At the same time, movement of vast quantities of European goods flowing from Tampico, Brownsville, and Matamoros moved northward into Texas while tons of Southern cotton were being shipped in the other direction; bound for European markets.

Two years after the Confederacy collapsed in the spring of 1865, the French were forced to leave Mexico. Forces loyal to Juarez quickly tightened the noose and drove Vidaurri and Maximilian from Mexico City to their last stand at Queretaro. When Maximilian was captured, Vidaurri attempted to escape but was caught and executed by firing squad in the town square of Santo Domingo on July 8, 1865, without the benefit of a trial.

Ignacio Sequin Zaragoza

March 29, 1829 – September 8, 1862

General Ignacio Zaragoza was born in the town of Presidio de la Espiritu Santo (also known as Goliad, Texas), in what was then the Mexican state of Coahuila. After his family moved to Monterey in 1844, Ignacio entered the seminary, but when the United States invaded Mexico in 1847, he volunteered to serve in the army but was rejected for military service. Consequently he joined the militia of Nuevo Leon as a sergeant. In 1855 his regiment was incorporated into the regular Mexican Army and Zaragoza was promoted to captain. During the 1850s he fought with the liberal forces and took part in the battles at Saltillo and Monterey against the armies of Antonio Lopez de Santa Anna.

During the War of Reform (1857-1860) he sided with Benito Juarez, taking part in several engagements including the Battle of Guadalajara and Calpulalpan in defense of the reformists' principles and democracy. In April 1861 Juarez promoted Zaragoza to Minister of War. However, the impending invasion of the French Army and Navy at Vera Cruz motivated him to resign and take personal charge of the Army of the East. Tragically, his wife died in Mexico City only a month before he began the construction of the Mexican defensive positions at Puebla in February 1862.

The French Army, trusting in their reputation as one of the most successful military organizations in the world, attacked the defenses of Puebla On May 5 expecting to walk over the disparate collection of Mexican defenders. However fate, luck, and carelessness on the part of French commanders in the field all conspired to hand the French Army its first defeat since Waterloo. Zaragoza's forces drove the French from the battlefield and back to the coast. It was a temporary setback for the French, but nonetheless a humiliating defeat at the hands of what was considered a ragtag army of Mexican peasants and Indians. In reality a good portion of the Mexican Army was well armed and trained. During a counter charge, which the French could not repel, over 1,000 of Napoleon's troops were killed or wounded. Although the French took Mexico City the next summer, the victory at Puebla is still considered one of the greatest military victories in Mexican history.

During on-going victory celebrations in Mexico City, Zaragoza contracted typhoid fever and died on September 8, 1862.

FRENCH PARTICIPANTS

Francois Achille Bazaine

February 13, 1811 – September 23, 1888

General de Division Francoiz (Marshal) Bazaine was born at Versailles, France. He enlisted in the French Army as a private in 1831 and served in Algeria in 1833 receiving the commission of sub-lieutenant in the Foreign Legion and rose rapidly through the ranks. Bazaine was promoted to lieutenant in 1835 and captain in 1837. Fighting in Algiers against Carlist guerillas in Spain he was awarded the first of many medals for valor, which would include the Cross (Chevalier) of the Legion d'Honneur. Bazaine participated in the second campaign with the Legion in Spain in 1837-38 and was wounded in the leg at the Battle of Barbastro. In 1844 he was wounded again, shot through the wrist at the Battle of Macta. He was rapidly promoted over the next four years to major and then to lieutenant colonel in 1848. Bazaine became a full colonel in 1850 and was given command of the First Regiment of the Foreign Legion headquartered in North Africa. He was promoted to brigadier general at Gallipoli and fought several battles during the Crimean War commanding the left wing of the French forces' final assault on Sebastopol on September 8, 1855. Wounded once again at Sebastopol by a shell fragment which struck him in the left hip, Bazaine received high acclaim from the Allied Command and was promoted to major general on September 22, 1855.

In 1862, Bazaine and the French Foreign Legion were sent to Mexico where he commanded with great distinction under General Forey, defeating the Mexican forces at the second Battle of Puebla and driving the Mexican president, Benito Juarez, into the northern frontier of his own country. Bazaine fought the Mexican rebels (what was left of any organized Mexican Army was dispersed throughout the land), and established a semblance of control around the main cities of the country. However, the Mexicans fought a tenacious guerilla war constantly attacking French supply lines and railroads. Eventually the threat of U.S. intervention, fate, along with the complication of Bazaine's marriage to a rich Mexican lady whose family was known as staunch supporters of Benito Juarez's movement, all conspired to end Napoleon III and Emperor Maximilian's grand adventure in Mexico.

When the French Army left Mexico for good in 1867, upon his return to Paris, Bazaine was appointed Commander-in-Chief of the Imperial Guard and awarded a seat in the Senate as a Marshal. When the Franco-Prussian War broke out in 1870, Bazaine was back in the forefront leading the forces of the Third Corps of the French Army of the Rhine, but his fortunes were about to take a tragic turn. The Prussian Army was known to be an excellent military machine with a reputation for its speed and menacing efficiency. In the first few months of the war the French suffered several defeats and Napoleon chose Bazaine to take chief command of the Army. Expected to immediately turn things around and save France from the hands of the Prussian forces, Bazaine reluctantly took charge of a situation he already believed to be lost. Joining his command in the field he found them ill equipped and numerically at a great disadvantage. Ordering his army into retreat back to Metz, Bazaine was wounded on August 14, 1870, by a shell fragment that struck him on the left shoulder. Making several tactical mistakes in the battles of Mars-la-Tour and Gravelotte, Bazaine's command found itself surrounded by the Prussian Army. On October 27, 1870, Bazaine surrendered his army of 140,000 troops to Otto von Bismarck, the Prussian Chancellor. For this action Bazaine was sentenced to death; however, his sentence was commuted to twenty years in prison. A year after his incarceration, Marshal Bazaine escaped from prison on August 9, 1873. He died five years later on September 23, 1888 while living in Spain.

Charles Louis Napoleon Bonaparte III

April 20, 1808 – January 9, 1873

Popularly known both as Louis Napoleon and Napoleon III, he was the son of Louis Bonaparte, the brother of Napoleon I. Following his uncle's defeat at Waterloo, Belgium in 1815, the Bourbons returned to power in France and the entire Bonaparte family was forced into exile. Young Louis was raised and educated in Switzerland and Germany, and as a young man settled for a while in Italy where he and his brother gravitated toward liberal politics.

Louis Napoleon was drawn into the Bonapartist movement, which attempted to restore the Bonaparte family to power and France to the glory it had known during the days of Napoleon I. He secretly returned to France in October 1836 to lead a failed coup that resulted in his exile to the United States where he spent the next four years living in New York City. Then he again returned secretly to France but was soon captured and sentenced to life in prison.

It was during his imprisonment at Ham that he wrote several essays and pamphlets that came to define Bonapartism as combining the monarchial claim with progressive social economic proposals that appealed to a broad segment of the French working classes. A sympathetic workman in the prison, perhaps inspired by these ideals, helped Louis Napoleon to escape to Southport, England.

Napoleon participated in the French presidential elections and won in a landslide victory on December 10, 1848. His winning edge was due to the support of the industrial working class who were not only impressed by his progressive economical views, but probably also the popular memory of his uncle, Napoleon I, and the great era of stability and military greatness he engendered.

Very quickly a series of successes followed Louis Napoleon's election. France was modernized with increased mining and factory operations and vast renovations of the city slum areas, and the expansion of roads and the railroad system. Militarily France formed an alliance with Britain and fought in the Crimean War (March 1854–March 1856). He also introduced the Vietnamese to European colonialism when a French expeditionary naval force bombarded Saigon in 1858, resulting in the cession of three southern Vietnamese provinces to France. Next a French protectorate was established in Cambodia expanding French power even further in Indo China. During the same period France participated in the Second Opium War that led to China's conceding to the French wide trading rights, free navigation of the Yangtze River, and freedom of religion for Christians in China.

The American Civil War and instability in Mexico gave Bonaparte an opportunity to attempt the fulfillment of a decades old dream of reasserting French influence in the western hemisphere. He further demonstrated his adventuristic foreign policy when he sent another expeditionary force to Mexico in 1862. Having expressed sympathy for the Confederate cause and hoping that a Southern victory would divide the United States into two less powerful nations, Louis Napoleon convinced Britain and Spain to join him in his Mexican scheme. Almost immediately the British and Spanish withdrew, seeing the adventure for what it was: an obvious attempt on France's part to establish a permanent presence in Mexico and the western hemisphere.

Without British/Spanish support the plan was doomed. Even though Mexican conservatives and the Church supported Bonaparte, he was still unable to win complete control of Mexico. Now, in 1863 he turned to another gambit: legitimize the French intervention by installing the Hapsburg prince of Austria, Maximilian, to reign as Emperor of Mexico.

Under Maximilian the French were relatively successful until the spring of 1865 when the Confederacy surrendered. Soon after, the United States Government began increasing support of the Benito Juarez government. That support came in a multitude of ways from military and political, to financial. Soon after the Confederate defeat, the Andrew Johnson administration invoked the Monroe Doctrine as a justification to force the French to withdraw from Mexico, which they did in 1867. In what proved to be a misguided sense of loyalty and honor, Maximilian stayed behind to face betrayal and finally execution at the hands of the Juaristas.

Louis Napoleon's troubles quickly multiplied after that. His health began to deteriorate and he was facing war with the emergent German Empire. The ensuing Franco-Prussian War humiliated France and ended Napoleon's dreams of a resurgent French Empire in Europe. He was captured by the Prussians at the Battle of Sedan in July 1870, and soon after went into exile. A haunted and bitter man he died of kidney failure in England on January 9, 1873. His most interesting distinction is that he was the last president of the French Republic, having been elected to that office, and also the last monarch of France as Emperor of the last French Empire. Louis Napoleon is buried (along with his son who died in 1879 while fighting the Zulus in Africa) at the Imperial Crypt in St. Michael's Abbey, Farnborough Hampshire, England.

Felix-Charles Douahy

August 14, 1816 – May 5, 1871

General Felix Douay.

His father enlisted Felix Douay in the French Marines in 1832, at the tender age off sixteen. He saw his first actions at Martinique and Guadeloupe in the French West Indies. His teachers and others considered Douay extremely intelligent, and by 1838 he had received a commission and was quickly promoted through the officer ranks to full colonel. Douay participated in the fighting during the wars in Algeria and Africa from March 18, 1845 to June 27, 1848, and fought against Giuseppe Garibaldi's Garibaldiennes. He also fought in the Crimean War from 1854 to 1855.

Within weeks of Douay's marriage he was promoted to brigadier general on June 10, 1859, and sent to fight in the Italian campaign.

Given command of several divisions, Douay was deployed to Mexico on March 15, 1862. He fought the Juaristas from isolated posts in Northern Mexico for nearly five years. Douay was promoted to major general in January 1863.

In 1865 he became acquainted with General Jo Shelby at his command post in San Luis Potosi. Shelby had refused to surrender at the end of the American Civil War and led 750 men deep into Mexico, passing through San Luis Potosi on his way to Mexico City and meeting with the Emperor. Douay was impressed by the rebel cavalryman and actually offered him a command in the Imperial Army. Like General Jeanningros, he wanted the Confederates to fight for Emperor Maximilian and Napoleon. However Marshal Bazaine and Maximilian thought using the ex-rebel general and his cavalrymen would provoke the United States into taking some sort of action and consequently turned down Shelby and Douay's offer.

Later in March 1866 Douay gave Shelby temporary command of a company of French Cuirassiers with the same authority as a regular French officer and dispatched him to rescue and reinforce Colonel DuPreuil's command at Censola, which was under attack at the time by Juarista troops. Due to Douay's controversial action, Shelby was most likely the first Confederate general to become, even if only temporarily, a ranking officer in the French Army and Maximilian's Empire.

Upon his return to France he became aide-de-camp to the emperor from April 4, 1868 to July 17 1870. Captured at Metz by Bismarck's Army of the Rhine, he was a P.O.W. for six months. When released Douay returned to command the 4th Army Corps and became Inspector General of the French Army. He died on May 5, 1871, the ninth anniversary of the French defeat at Puebla. The cause of death was an unknown infectious disease passed to Douay by his wife who had died three weeks earlier from the same disease.

Elie Frederic Forey

January 10, 1804 – June 20, 1872

Marshal Elie F. Forey was born in Paris and attended the French military academy at Saint-Cyr, later accepting a commission as lieutenant in the Second Infantry Regiment in 1824. He saw his first action in Algiers in 1830 and was promoted to captain in 1835. By 1839 he was a colonel in command of a battalion of chasseurs a pied (foot soldiers). In 1852, having supported Napoleon III during the coup d'etat he was promoted to general de division.

Forey served in the Crimean War and commanded a division during the siege of Sebastopol and later in the Austro-Sardinian War of 1859. Forey commanded a division at the Battle of Montebello and later at Solferino, where Forey's troops distinguished themselves and their commanding officer by taking the last Austrian positions of the village of Cavriana.

In May 1862 Forey was appointed to command of the French expeditionary corps to Mexico. His army landed at Vera Cruz in September 1862, just four months after General Laurencez' stunning defeat at the first Battle of Puebla (ever after celebrated in Mexico as *Cinco de Mayo*). With Forey's reinforcements, the French Army moved inland from the Bay of Campeche while their navy bombarded the port of Vera Cruz, and defeated the Mexicans at the second Battle of Puebla.

From there he marched on and captured Mexico City, forcing Benito Juarez to flee to northern Mexico. He was awarded the Marshal's baton for this achievement and after participating in solidifying the French command in Mexico he turned over his command to Bazain and returned to France. Upon his arrival he took command of the Second Corps.

Suffering from a blood clot in his brain in 1867 Forey was put on the non-active list and, therefore, did not participate in the Franco-Prussian War. Forey died in Paris, the city of his birth in 1872.

Pierre Joseph Jeanningros

? 1816 – April 30, 1902

General Pierre J. Jeanningros was born in Saint-Ferjeux, France. At the age of seventeen he joined the army and initially served with the 666th Infantry Regiment de Ligne. On July 6, 1835 he was promoted to corporal and by April 21, 1836 became a fourrier grenadier. Before the year was out he became a sergeant in the Zouave Regiment. Rising through the ranks, Jeanningros became a lieutenant colonel by October 2, 1855. In the spring of 1863 he was given command of an expedition to Mexico to reinforce General Elie Frederic Forey and landed at Vera Cruz in April 1863.

In Mexico, Jeanningros was under the direct command of Marshal Bazaine and fought to bring Mexico under French military rule. From June 1862 to June 1867 he was the military commander of the states of Cacahueil, Leon, Queretaro, and Sierra, all in the wilds of northern Mexico where the Juaristas at the time were most active.

Jeanningros met with Confederate General Jo Shelby when he and his expeditionary force were stationed in Monterey. After some tense opening negotiations as to whether Jeanningros and Shelby were going to go to war with one another; Jeanningros rolled out the red carpet for his new friends, the Confederate exiles. After a day or two of meetings Jeanningros allowed Shelby to march his column to Mazatlan on the coast for a base of operations for the Confederate Rebels to organize and raise a military force of volunteers from the Southern American states who would be willing to fight for France. However, Marshal Bazaine overruled Jeanningros' offer and diverted Shelby's expedition to Mexico City to make his proposal directly to Emperor. Maximilian did not like the political implications of employing Confederate Army exiles and turned down the Jeanningros/Shelby idea of raising an American force to fight for the French Empire.

Colonel Jeanningros commanded over 5,000 troops in Northern Mexico of which were four regiments of the Foreign Legion. His various military camps scattered around Monterey and Parras were Maximilian's front line in the war against the Juaristas. He was also the commanding officer of the famous unit of Legionnaires who fought a savage last stand at Camerón on April 30, 1863 that has become a celebrated legend in the Foreign Legion yet today.

Then Colonel Jeanningros received orders to send two companies of Legionnaires to Chiquihuite to escort a convoy to Palo Verde where another French outpost awaited supplies. Jeanningros ordered Captain Jean Danjou with a company of sixty-two Legionnaires and two lieutenants to accompany the wagon and mule train of supplies to Palo Verde. However, near Camerón the French column came under attack by over 2,200 Mexican infantry and cavalry (three battalions). Forced to make a running fight for several hours and an eventual last stand at Hacienda Camerón the Legionnaires fought to the finish. According to the survivors' reports of the battle, after a day and a half with no water some of the wounded Legionnaires drank their own blood and urine as they pressed themselves against the adobe walls of the Hacienda to avoid the relentless Mexican fusillade of bullets, exploding cannon shells and cavalry charges. With less than a dozen men left and Captain Danjou dead; the remaining

Legionnaires chose to make a suicidal bayonet charge rather than surrender or wait for the Mexicans to decide how they would die. Charging into a sea of Mexican troops the Legionnaires were soon cut down with only three left standing back-to-back over their fallen comrades and surrounded by Mexican bayonets. The last French survivors were spared by the Mexican troops out of an almost superstitious respect. When one Mexican trooper was asked by his commanding officer why they did not kill the last five Legionnaires he stated, "Pero non son hombres, son demonios!" ("Because they are not men, but devils.")

On recommendation of Jeanningros, Napoleon III ordered the word "Camerone" with the names of Captain Danjou and Lieutenants Vilain and Maudet inscribed in gold on the walls of the Les Invalides in Paris. Returning to France in 1867, Jeanningros participated in the Franco-Prussian War and was taken prisoner on October 23, 1870 after Marshal Bazaine surrendered at Metz. Returning to France on March 12, 1871 he went on to have an illustrious career and retired in 1889. He died in Paris on April 30, 1902.

His Imperial and Royal Highness Ferdinand Maximilian, Joseph Ferdinand Maximilian, Joseph, Prince Imperial and Archduke of Austria, Prince Royal of Hungary and Bohemia

Maximilian I, Emperor of Mexico

July 6, 1832 – June 19, 1867

Maximilian von Hapsburg was born in Schonbrunn, Vienna, Austria, second in line to the Austrian throne. He was considered extremely intelligent and displayed an early interest in botany and science. Trained for the navy he rose to high command in quick fashion and helped create the naval port of Trieste on the Adriatic Sea and was instrumental in raising the fleet with which Admiral Wilhelm von Tegetthoff won several sea battles in the Italian War. His progressive ideas and reputation as an intellectual liberal won him his appointment as Viceroy of the Kingdom of Lombardy-Venetia. He married Princess Charlotte of Belgium (later to be known as the Empress Carlota of Mexico) on July 26, 1856 in Brussels, Belgium. One of the first Confederate colonies established near Vera Cruz (Carlota) was named after the Empress.

In 1859 Mexican monarchists with proposals to make him Emperor of Mexico approached Maximilian. Though probably intrigued by the idea, he resisted such attempts until Louis Napoleon of France intervened in Mexico and joined the monarchists. With the French intervention and General Elie Frederic Forey's capture of Mexico City, Napoleon saw Maximilian's acceptance of the throne of Mexico as imperative to his country's own success in Mexico.

Maximilian finally agreed if "the people of Mexico" accepted him as their emperor. After a hastily organized plebiscite seemed to indicate general approval, he consented and accepted the crown, landing in Vera Cruz on May 28, 1864 with the backing of the "Mexican Conservatore Movement" and Napoleon III. With a civil war in progress in the United States and another one in Mexico at the same time, Maximilian soon found himself a pawn in political forces beyond his or any one individual's control. Choosing to rule from Mexico City he selected as his palace Chapultepec Castle atop a hill that had once been the site of a retreat for Aztec emperors.

Maximilian attempted to bring Mexico under some sort of civil control through military force but was never successful at subduing Benito Juarez and his guerilla army. His liberal policies conflicted with the French military leadership and their brutal strategies for bringing Juarez and his followers to their knees. And, of course, the underlying question always was: how long would the U.S. allow the French to thumb their noses at the Monroe Doctrine?

The answer came when Lee surrendered on April 9, 1865 at Appomattox Courthouse. Soon after, the United States began to provide arms and other supplies to the republican-rebel forces under Benito Juarez, and to move to the Mexican border a heavy concentration of American troops, veterans of the recent American Civil War. To observers outside of Mexico, Maximilian's demise was obviously imminent. Under pressure from the American government Napoleon III withdrew his troops back to France, leaving Maximilian (along with the Confederates who had colonized in Mexico) vulnerable and exposed to their enemies. Napoleon could argue that he left Mexico, not because of American military pressure, but because of a looming war with Prussia. Napoleon urged Maximilian to flee from Mexico, but the Prince of Hapsburg was truly loyal and dedicated to the Mexican people and his followers and considered it his duty to stay and face his fate with them.

Forced to abandon Mexico City Maximilian withdrew to Queretaro in February in 1867 and was soon under siege. The city fell on May 15, 1867 and Maximilian was captured, court-martialed and sentenced to die by firing squad. The sentence was carried out on June 19, 1867. It was said that Maximilian bribed all seven riflemen in the firing squad not to shoot him in the head, but one of his executioners did so anyway.

The remains of Maximilian I, Emperor of Mexico, were removed from Mexico in 1868 and buried in the Imperial Crypt in Vienna, Austria. Maximilian's older brother reigned as Franz Joseph, Emperor of the Austro-Hungarian Empire through World War I (1914-1918).

CONFEDERATE PARTICIPANTS

Henry Watkins Allen

April 29, 1820 – April 22, 1866

General Henry Watkins Allen was an American soldier, politician, journalist, writer, teacher, lawyer and adventurer. He was born in Prince Edward County, Virginia, and educated at Marion College, Missouri, and studied law at Harvard. By the time the Civil War began Allen had acquired an assortment of arguably unorthodox military experiences having joined the Texas Revolution and traveling across Europe intending to join the Italian struggle for independence.

His most serious military experience came with the opening of the Civil War. He entered the Confederate Army as a lieutenant colonel in 1861, and participated in the Battle of Shiloh where he received a devastating facial wound. Shot through both cheeks and with blood gushing from his mouth, Allen stuffed handfuls of cotton into his mouth and tied a handkerchief abound his face and then continued to lead his men until he nearly lost consciousness from loss of blood. Demonstrating incredible toughness, he was back in the saddle the next morning, in the thick of the fight and rallying his troops.

Four months later Allen was in action at the Battle of Baton Rouge on August 5, 1862, when he was severely wounded and permanently crippled when he was hit in both legs by cannon-fired canister shot. His left leg was slightly wounded but his right leg was shattered. He refused to allow the surgeons to amputate the leg and, contrary to their medical opinions, Allen miraculously survived; but he was never the same physically, though he was promoted to brigadier general a year later.

He had just been appointed brigadier general when he was elected to the office of Governor of Louisiana in 1864. Allen governed Louisiana with dictatorial measures as he addressed the dire situation in his state as a result of the war. He organized the state's stores of commissary, oversaw the building of foundries and factories and is credited with saving west and north Louisiana from starvation and lawlessness in the waning days of the war. After circumventing the Union blockade and restoring the cotton exports through Texas to Mexico, he also authorized a geological survey of Louisiana to locate raw materials and authorized several mining operations.

When the Confederacy collapsed, Allen supported General Kirby Smith and others, who refused to surrender and went into exile in Mexico in June 1865. With John Newman Edwards he began an English-language newspaper called The *Mexican Times* in Mexico City that chronicled the Confederate exile experience in Mexico. While living in Mexico Allen also wrote a memoir of his earlier travels in Europe called *Travels of a Sugar Planter*.

While the Confederate colonization movement in Mexico was still showing promise of success Allen, who had never completely recovered from his war wounds, died on April 22, 1866, at the age of forty-five. He was temporarily buried in the still extant American Cemetery in Mexico City. In January 1876, his remains were disinterred and returned to New Orleans, then to Baton Rouge where he was later buried in the Old State Capital.

Hamilton Prioleau Bee

July 22, 1822 – October 3, 1897

Brigadier General Hamilton P. Bee was born in Charleston, South Carolina, however his family moved to Texas when he was fourteen. In 1839, at the age of 17, he served as secretary for the commission that established the boundary between the Republic of Texas and the United States. In 1843 President Sam Houston sent Bee on a mission to convene a peace council with the Comanches, which he accomplished by August 1843 with the Treaty of Tehuacana Creek. He was named secretary of the Texas Senate in 1846.

By the time Mexico and the United States went to war in 1846, Bee was well established on the Texas scene. He enlisted as a private in the cavalry and rose to the rank of lieutenant by war's end. After the war he moved to Laredo where he ran for and was elected to the Texas Legislature and served from 1849 to 1859, acting as Speaker of the House from 1855 to 1857.

When the Civil War began in 1861, Bee was elected brigadier general of the state militia and was later appointed to the same rank in the Confederate Army on March 4, 1862. He was also appointed to command of the lower Rio Grande District, which was notorious for its lawlessness. Also in his area of responsibility were strategic ports on the Gulf coast through which came munitions and other supplies smuggled through the Union blockade.

As commander of the West Sub-District of Texas Bee and Albino Lopez, the Juarista Commandant of Tamaulipas, negotiated an agreement in February 1863, which allowed for the enforcement of law on both sides of the Rio Grande. Later that year he was credited with saving millions of dollars worth of munitions, cotton and other Confederate stores at Brownsville when he repulsed a Union force led by General Nathaniel P. Banks. In the spring of 1864 he was transferred to a field command in the Red River Campaign under Lieutenant General Richard Taylor. Bee was wounded in the face leading a cavalry charge at Pleasant Hill, Louisiana on April 9, 1864. In that same engagement he had two horses shot from under him and still returned to the field the next day.

Following the war Bee joined the Confederate exodus to Mexico and stayed there much longer than most of the Southern Exiles, not returning to Texas until 1876 when his eyesight began to fail. Over the next few years his general health declined along with his eyesight. His death on October 3, 1897 was attributed to "disease of the heart." Bee is buried in the Confederate Cemetery in San Antonio.

Jubal Anderson Early

November 3, 1816 – March 2, 1894

Lieutenant General Jubal A. Early was born in Franklin County, Virginia, and graduated from the United States Military Academy in 1837. The next year he was fighting in Florida against the Seminoles as a second lieutenant in the 3rd U.S. Artillery regiment. He left the Army for a brief career as prosecuting attorney for Franklin and Floyd Counties in Virginia, but duty called in 1846 when war began with Mexico, and he rejoined the Army.

Early joined the Confederate Army in 1861 and was quickly promoted to brigadier general for his performance leading a brigade at First Manassas. Throughout the war he fought in most of the major engagements in the east. His rough demeanor and short temper led Lee to refer to Early as his "Bad Old Man."

After Lee's surrender Early fled to Texas hoping to find Confederate forces still holding out in the west that he could join up with and continue the war. Frustrated when Kirby Smith surrendered the Trans-Mississippi Department to the Yankees he fled to Mexico. Leaving Mexico to its troubles with the collapse of the colonization movement there he traveled to Cuba and Canada where he wrote his memoirs as well as several articles for the

Southern Historical Society which helped to establish the cultural phenomenon referred to as the "Lost Cause Myth." He eventually returned to Virginia but the bitterness never left him.

General Early died in 1894 in Lynchburg from injuries sustained in a fall down a flight of stairs. He is buried at Spring Hill Cemetery.

John Newman Edwards

January 4, 1839 – May 4, 1889

Major John Newman Edwards was born in Warren County, Virginia, where as a boy he learned typesetting and showed an early literary talent. Edwards joined General Jo Shelby's Cavalry Regiment near Waverly, Lafayette County, Missouri in 1862 and was soon appointed Brigade Adjutant with the rank of major. He fought along side Shelby at the Battles of Prairie Grove, Cane Hill, Hartsville, Boonville, and Newtonia, and was wounded and captured during the Cape Girardeau Expedition. Later Edwards was exchanged to rejoin Shelby just before the disastrous assault at Helena, Arkansas.

At the end of the war Edwards joined Shelby in exile in Mexico. He wrote *Shelby's Expedition to Mexico, An Unwritten Leaf of the War*, the only extensive, contemporary account of the trek of Shelby's brigade from Texas to Mexico City. Preceding Shelby to Mexico City Edwards managed to obtain a grant of land upon which was established the Confederate Colony at Cordoba. Later he joined Governor Henry Watkins Allen in establishing The *Mexican Times*, an English-language newspaper catering to the Confederate expatriates in Mexico.

Edwards returned to Missouri in 1867 and founded The *Kansas City Times* continuing a life-long and prominent career in journalism and literature. He maintained his close relationship with Jo Shelby and other friends from the war and the Mexican expedition, even appearing in court as a character witness for Frank James.

He also published several books and journals on the bloody border war between Kansas and Missouri, the Civil War in Missouri and Arkansas, and his post-war experiences in Mexico.

He died suddenly and unexpectedly at the McCarty house in Jefferson City on Saturday morning, May 4, 1889. His death was attributed to cardiac disease.

John Salmon Ford

May 26, 1815 – November 3, 1897

Colonel John S. Ford was born in Greenfield District, South Carolina but his family soon moved to San Augustine, Texas. As a young man Ford served from 1836 to 1838 in the Texas Army. He was one of the most politically active individuals to emerge from the lower Rio Grande area in the mid-to-late 19th century. He became known as a prominent military figure as well as a respected newspaper editor and congressman. Fighting Apaches, Mexicans, Yankees, and renegades up and down the Rio Grande, Ford participated in most of the military struggles in Texas over a fifty-year period and lived to relate his experiences through his written memoirs.

Ford was elected in 1844 to a seat in the Texas Congress and became the editor of The *Texas Democrat* in Austin. A "renaissance man" of many talents Ford also studied medicine as well as law and practiced both in Austin for several years. During the Mexican-American War Ford commanded a company of spies operating on both sides of the border. He was commended for recognition by General Joseph Lane and received honors for "gallant service."

In 1849 Ford and Robert S. Neighbors led an exploratory expedition of the wild country between San Antonio and El Paso. He later published an essay and map entitled *The Ford and Neighbors Trail*. From 1849 to 1852, when he was elected to the Texas Senate, Ford was stationed along the Rio Grande near Nueces, fighting Indian Tribes who were attacking the local citizenry.

Ford was dispatched south of the Rio Grande in 1859 to find and destroy Juan Cortina's renegade army that was terrorizing the area around Brownsville and Matamoros. In late December Ford led a company of Texas Rangers in a series of engagements against Cortina that drove the bandit into the mountains and temporarily dispersed his army.

During the Civil War, Ford led the Second Texas Cavalry and continued to fight Indians, Mexicans, and Union troops; and protecting Confederate supply lines from 1863 to the end of the war. Ford participated in the Confederate victory at Palmito Ranch near Brownsville, Texas, on May 5, 1865, over a month following Robert E. Lee's surrender to Grant at Appomattox.

Ford became mayor of Brownsville in 1873 and continued to edit his newspaper, The *Brownsville Sentinel*. Ford's memoirs are in the custody of the Texas State Historical Association, of which he was a charter member. He died in San Antonio on November 3, 1897.

William Polk Hardeman

November 4, 1816 – April 8, 1898

Brigadier General William P. Hardeman was born in Williams County, Tennessee on November 4, 1816 and briefly attended the University of Nashville before moving to Matagorda County, Texas in 1835 at the age of eighteen. He joined the Texas revolution in 1836 and with a small band led by Philip Dimitt he marched off in response to William Travis's call for help at the Alamo. They ran into a group of Mexican pickets near Gonzalez and were turned back. From then until the beginning of the Civil War he was almost constantly engaged in fighting Comanches and Apaches as a Texas Ranger and private citizen.

When the Civil War broke out Hardeman raised a company of men from Guadelupe and Caldwell Counties and joined General Sibley's New Mexico Campaign, which was aimed at taking what is now Arizona and New Mexico as a Confederate territory. He subsequently fought from Albuquerque to Louisiana, and participated in the Red River Campaign under General Richard Taylor.

After the war Hardeman joined the exodus of former Confederates to exile in Mexico. He attempted to establish a colony near Guadalajara while serving as a land surveyor for

Maximilian's government and acted as a settlement agent for Confederate colonists.

As the colonization movement began to crumble Hardeman returned to Texas in 1866. His service there hardly missed a beat for the next thirty years, as he was involved in endeavors ranging from Inspector of Railroads to education. He helped to establish the Agricultural and Mechanical College of Texas, now known as Texas A&M University. "Old Gotch," as Hardeman was called, holds the distinction of having served Texas in either a military or governmental position longer than any other Texan in history.

Hardeman died of Bright's disease on April 8, 1898, and rests in the Texas State Cemetery in Austin.

Isham Green Harris

February 10, 1818 – July 8, 1897

Governor Isham G. Harris was born near Tullahoma, Tennessee, and was educated at Winchester Academy in Winchester, Tennessee, He moved to Paris, Tennessee and began his study of the law later opening his first practice in his hometown of Paris. He was elected to the Tennessee State Senate in 1847 and served one term before being elected to the United States House of Representatives. Harris was elected Governor of Tennessee in 1857, succeeding Andrew Johnson, who remained loyal to the Union and eventually became Lincoln's vice president and successor as president.

When war came in 1861, Harris refused to answer President Lincoln's call for 75,000 volunteers stating, "Tennessee will not furnish a single man for the purpose of coercion, 50,000 if necessary for the defense of our rights and those of our Southern brothers." When Lincoln appointed Andrew Johnson military governor of Tennessee, Harris faced a dual governorship: one elected by the people of Tennessee, the other federally appointed. Although Harris did not formally resign the governorship, he did become staff officer in the Confederate Army for General Albert Sidney Johnston and later served under General Joseph E. Johnston.

Following Lee's surrender, Harris fled the country ahead of a $5,000 bounty for his capture. He traveled to Mexico with General Hylan B. Lyon and other Confederates hoping to ally himself with Maximilian and to help in building a new Confederate community in Mexico. When the colonization movement fell apart with Maximilian's execution and the departure of the French Army, Harris abandoned Mexico and sought temporary refuge in England.

Returning to Tennessee in 1868, he resumed his law practice in Memphis and was subsequently elected to four terms in the United States Senate. Harris died on July 8, 1897, and his funeral was held in the Senate chamber of the United States Capitol. Governor Harris is buried alongside scores of Confederate luminaries at Elmwood Cemetery in Memphis.

Thomas Carmichael Hindman, Jr.
January 28, 1828 – September 28, 1868

Major General Thomas Hindman was born in Knoxville, Tennessee and later moved with his family to Ripley, Mississippi where he attended the Lawrenceville School and graduated with honors on September 25, 1843. During the Mexican-American War he served in Tippah County with the 2nd Mississippi Regiment as a lieutenant and later as a captain. He and his good friend Patrick Cleburne were wounded in a street gun battle with a rival political group in 1856. The surgeon cut out a bullet lodged behind his pectoral muscle while Hindman sat "calmly smoking a cigar." He served as a member of the Mississippi House of Representatives from 1854 to 1856 and was elected to the U.S. Congress from the 1st Congressional District of Arkansas. That same year he was permanently disabled when a tree blew over in a storm and smashed into his carriage, injuring his left leg.

When Arkansas seceded from the Union in May 1861 Hindman immediately joined the Confederate Army and raised ten companies of men for the service at Helena. During the war he proved himself to be a fearless and aggressive leader and suffered wounds in several major battles from Shiloh, after which he was promoted to brigadier general, to Kennesaw Mountain, where during the fighting he was struck in the face by a tree limb, which caused serious impairment to his vision for several months taking him out of the fighting for the rest of the war.

Upon the surrender of the Confederate Trans-Mississippi Department Hindman joined the exodus to Mexico where he attempted to practice law and become a coffee planter in the town of Carlota. Still in Mexico when Maximilian was executed he petitioned President Johnson for a pardon but was denied. He returned to Helena, Arkansas in 1868 anyway and resumed his law practice.

Hindman became a leader of the "Young Democracy," a new political movement attempting reconciliation with the Union. His activities during this period were controversial and earned him enemies on both sides of the stands he took.

On September 27, 1868 Hindman was sitting in the parlor or his home in Helena with his children nearby when one or more assailants fired through the window striking him in the throat, chest, and hand. Though weak from loss of blood he made his way to his front porch to address a gathering of neighbors who had come to express their concern. He died several hours after being shot. His killer or killers were never found, nor is their motive known.

Hindman was laid to rest in the Maple Hill Cemetery (known at the time as the Evergreen Cemetery) near his friend Patrick Cleburne.

Hylan Benton Lyon

February 22, 1836 – April 25, 1907

Brigadier General Hylan B. Lyon was born in Lyon County, Kentucky into a wealthy plantation family. He inherited the family estate when both his parents died while he was still a child. Having studied at Cumberland College and the University of Kentucky Lyon received an appointment to the United States Military Academy at the age of sixteen and went on to graduate nineteenth in the class of 1856. He was brevetted a second lieutenant in the Second United States Artillery and fought in the Second Seminole War. Later he was transferred to Fort Hyuma, California and a year after that he was ordered to the Washington Territory where he fought several battles against local Indian tribes.

Lyon joined the Confederate Army in 1861 as a lieutenant colonel, and personally organized Lyon's Battery of the Eighth Kentucky Infantry. Taken prisoner with his battery at the fall of Fort Donelson he was later exchanged and went on to serve under General Earl Van Dorn and was later a participant in the siege of Vicksburg under the command of General John C. Pemberton. General Braxton Bragg later appointed Lyon to the command of two regiments of cavalry under "Fightin'" Joe Wheeler. Lyon holds the distinction of having fought in the Civil War as an infantryman, artillerist, and cavalryman, the only general officer in the war have done so.

In 1864 Lyon became a cavalry commander in Mississippi with the rank of brigadier general and rode with Nathan B. Forrest during some of Forrest's greatest victories, such as the battles at Okolona and Brice's Crossroads. In December Forrest sent Lyon to invade Kentucky and enforce the CSA draft law there and to create a diversion for Forrest's current operations. Before returning to Forrest's command twenty-three days later he had executed lightning raids that destroyed significant Union stores, equipment and supplies, and burned seven courthouses.

Forrest biographer and Confederate soldier John Wyeth described Lyon as "one of Forest's most devoted followers, a stubborn fighter, exhibiting at times a recklessness akin to desperation." An example of Lyon's "recklessness" is exemplified during an episode when he was temporarily taken prisoner while on a special mission for Forrest. While sleeping at the home of a Mr. Noble in Red Hill, Alabama, Lyon was surprised by a knock on his door and upon opening the door he found himself staring down the barrel of a pistol held by a Union soldier also named Lyon. General Lyon briefly capitulated but while dressing took advantage of a momentary distraction that allowed him to retrieve the two pistols he always kept under his pillow while sleeping. In a flash he killed his would-be captor and escaped.

When the war ended Lyon chose exile to surrender and accompanied Governor Isham G. Harris of Tennessee to Mexico City to seek a new life. In Mexico Lyon worked for over a year as a civil engineer under Maximilian, but when he learned that he would not be prosecuted for his activities in Kentucky during the war, he returned home.

Upon his return Lyon settled in Eddyville, Kentucky, where he resumed his farming and mercantile business and later served as commissioner of the Kentucky State Prison System. General Lyon died suddenly and unexpectedly on April 25, 1907, and is buried in the cemetery at Eddyville.

John Bankhead Magruder

May 1, 1807 – February 18, 1871

Major General John B. Magruder was born in Port Royal, Virginia. He began his higher education at the University of Virginia, and as a student there had the opportunity to dine with former president Thomas Jefferson. A military profession beckoned and Magruder left Virginia to become a cadet at West Point. Graduating in 1830 he embarked on an illustrious military career during which he served in three different armies: the United States Army, the Confederate Army, and the Imperial Mexican Army.

Known as "Prince John," Magruder was a flamboyant rogue, an actor and singer who spoke with a slight lisp, except when he was singing. He was said to have a very good tenor voice which he frequently employed in impromptu performances. Magruder loved to compose songs and stage concerts and amateur theater productions to relieve the tedium of peacetime military duty.

Any temptation to assume that "Prince John" was some sort of dandy should be readily dismissed, for on the battlefield he proved himself a brave and fearless soldier on numerous occasions. In the Second Seminole War in Florida he served with distinction, and later under Winfield Scott in Mexico he quickly rose in rank because of his military skill and daring.

During the Civil War Magruder fought against Union General George B. McClellan during the Battle of Yorktown delaying the Federal forces by deception and delaying McClellan's proposed march on Richmond for several weeks. Serving later in Texas he led the Confederate force that recaptured the port of Galveston for the Confederacy. From August 1864 to march 1865, Magruder commanded the Department of Arkansas and then returned to command the District of Texas, New Mexico, and Arizona until Kirby Smith surrendered the Army of the Trans-Mississippi.

Magruder fled to Mexico where he became a major general in Maximilian's imperial Army of Mexico. When Maximilian was captured and executed in 1867, he found himself fleeing in the other direction; back to Texas. After lecturing on Mexico in several mid-Atlantic cities he settled in Houston but fell into deteriorating health and died of heart disease on February 18, 1871.

"Prince John" was buried in Houston, but his remains were later returned to the scene of his most lasting military achievement, Galveston, where he now rests in the Episcopal Cemetery.

Matthew Fontaine Maury

January 14, 1806 – February 1, 1873

Matthew Fontaine Maury was born in 1806 in Spotsylvania County, Virginia. His family moved to Franklin, Tennessee in 1811. Family friend Senator Sam Houston helped him to receive a naval appointment as a United States Navy midshipman in 1825 when he was nineteen years old. Maury thus began his lifelong study of oceanography, naval meteorology, navigation, and the charting of wind and sea currents.

Maury's scientific interest in all things affecting the seafarer led to his publication of two major works: *Wind and Current Chart of the Atlantic and Sailing Directions*, and *Physical Geography of the Seas and its Meteorology*. These studies remain masterpieces studied to this day and Maury is still known as "The Pathfinder of the Seas," and "The Father of Modern Oceanography." A serious injury ended his seafaring days and resulted in his appointment as the first Superintendent of the U.S. Naval Observatory in 1842, a post he held until the beginning of the Civil War in 1861. He also advocated for a naval academy as a counterpart to U.S. Military Academy at West Point, which finally became a reality in 1845.

At the outbreak of the Civil War, Maury resigned his U.S. Navy commission to join the Confederate Navy. He was appointed

Chief of Sea Coast, River and Harbor defenses and set to work in earnest to fulfill his duties. One result was the perfection of the electric torpedo, based on the principle of the modern contact mine, which reputedly cost the Union more losses than from any other causes combined.

Maury went to Mexico after the war and became Imperial Commissioner of Immigration under Emperor Maximilian that put him in an advantageous position to assist displaced Confederates and with the foundation of the Carlota and New Virginia colonies. He returned to the United States after the fall of Maximilian and the Confederate colonies and accepted a position in the Physics Department at VMI and later advocated for the establishment of an agricultural college, which resulted in the founding in 1872 of the Virginia Agricultural and Mechanical College, known today as Virginia Tech in Blacksburg, Virginia. Offered the presidency of William and Mary University Maury declined so he could be near General Robert E. Lee, at whose funeral he was a pallbearer.

Declining health compounded by severe exhaustion from his final lecture tour in Europe, Commodore Maury died on February 1, 1873. He is buried in the Hollywood Cemetery in Richmond between Presidents James Monroe and John Tyler.

John McCausland, Jr.

September 13, 1836 – January 22, 1927

Brigadier General John McCausland, Jr. was born to Irish immigrants in St. Louis, Missouri on September 13, 1836. He began his higher education at the University of Virginia but transferred to the Virginia Military Institute after a year. He acted as assistant professor of mathematics at VMI while completing his military studies there. While McCausland was teaching at VMI he was selected to make up a military guard to protect John Brown during his trial and execution following Brown's raid on the Federal armory at Harper's Ferry in 1859.

When Virginia seceded from the Union in 1861, McCausland offered his services to his adopted state and was commissioned a lieutenant colonel on April 29. He immediately set about organizing the 36th Regiment of Virginia Infantry, which he led to Bowling Green, Kentucky, to join General Albert Sidney Johnston. McCausland fought at Fort Donelson and escaped with his command before the fort was surrendered to General Ulysses S. Grant in February 1862.

McCausland became a brigadier general on May 18, 1864, and while serving under

Early participated in the burning of Chambersburg, Pennsylvania, in reprisal for the earlier destruction of his alma mater, VMI. He fought in several subsequent battles and was wounded during the fighting at New Market and Milford. Prior to Lee's surrender to Grant McCausland was engaged in fighting with Union troops around Appomattox.

General McCausland was known for his unyielding spirit. As he had refused to surrender at Fort Donelson at the beginning of the war, he chose to escape the surrender at Appomattox with his cavalry. A short time later he disbanded his troops and, expecting to be hunted down for his part in the burning of Chambersburg, he went into exile. Traveling to Canada, England, France, and Scotland, McCausland finally joined a growing body of Southern exiles in Mexico where he remained until 1867.

Upon his return to the United States McCausland was charged with arson in Pennsylvania, but now-President Grant intervened and gave him a pardon. He settled on a large tract of land in West Virginia at Grimm's Landing. He did not get along well with his Unionist neighbors who continued to blame him for the razing of Chambersburg. This led him to eventually become very reclusive.

When McCausland died on January 22, 1927, at home on his farm he was the last of only two living Confederate generals.

Sterling Price

September 20, 1808 – September 29, 1867

Major General Sterling Price was born near Farmville, Prince Edward County, Virginia. He worked his way through his law studies at Hampden-Sydney College as a courthouse clerk. He moved to Fayette, Missouri in 1832, where he married Mary Head, the mother of their seven children.

During this period Missouri was in turmoil over the Mormon settlements in the western part of the state. In 1838 Price was a member of the military delegation sent to investigate the reported disturbances created by the Latter-day Saints near Independence. His report was favorable to the Mormons stating that he found them innocent of any infractions of the law.

Price was serving in the Missouri House of Representatives when war broke out with Mexico. He raised the Second Regiment, Missouri Volunteer Cavalry, and marched them to Santa Fe, New Mexico, where he assumed command of the entire territory. President Polk appointed him brigadier general of volunteers on July 20, 1847. He returned to Missouri after the war and bought a farm where he began raising tobacco, and became a slave owner.

Enjoying immense popularity Price was elected governor of Missouri (1853-1857).

Commissioned a major general in the Confederate Army, Price saw action on several battlefields. Twice wounded in battle, he refused to leave the field and continued to lead his men. He also developed dysentery, which plagued him for the rest of his life and was probably partly responsible for his continued declining health.

Despite his health Price joined the exodus to Mexico after the war and sought service under Maximilian. While making arrangements to take up residence in the Confederate colony of Carlota he contracted typhoid fever, which was complicated by his weight loss from dysentery. Within a year friends reported that they could hardly recognize him because of the ravages his illness.

Price returned to St. Louis in December 1866, but never regained his health. He died on September 29, 1867, and was buried in Bellefontaine Cemetery.

Thomas Caute Reynolds

October 11, 1821 – March 30, 1887

Thomas Caute Reynolds was born in Charleston, South Carolina. He had an excellent education graduating in 1838 from the University of Virginia and in 1842 he received a doctorate of law degree from the University of Heidelberg. Reynolds opened a law practice in St. Louis in 1850 and later served as United States Attorney. He was an early pro-slavery Democrat and established his commitment to the cause when he disagreed with an article written by Republican newspaperman Grantz Brown. The two fought a duel in which Brown was hit in the leg.

In 1860 Reynolds was elected Pro-Southern lieutenant governor of Missouri, but with the state effectively controlled by the Union he had little opportunity to govern. His only battle experience during the Civil War came during General Sterling Price and Kirby Smith's raid into Missouri in October 1864.

Governor Reynolds joined other Confederates in their march to Mexico when the Confederacy surrendered. He landed in Mexico City in the summer of 1865 and became railroad commissioner and unofficial advisor to Emperor Maximilian. When the Southern colonization movement failed he returned to St. Louis in 1869 and opened a new law practice. He was appointed by the Cleveland administration in 1874 to serve as trade commissioner to South America.

Reynolds was never able to fully rehabilitate himself after the war. Financially wrecked from his failing law practice and other unsuccessful business ventures, he climbed to the top of the tallest building in St. Louis, the eleven-story Customs House, and jumped down an elevator shaft to his death. On his body authorities found a note indicating that Reynolds believed he was suffering from a mental disease and losing his mind.

A newspaper headline on March 30, 1887 read: "A Fearful Death Plunge; Suicide of Ex-Gov. Thomas C. Reynolds. He Leaves A Letter In Which He States That Life Had Become a Burden To Him." His body rests in the Bellefontaine and Calvary Cemetery in St. Louis.

Joseph Orville Shelby

December 12, 1830 – February 13, 1897

Brigadier General Jo Shelby, descendant of an illustrious family in American history, was born in Lexington, Kentucky. His father died with he was three years old, and his mother later married a wealthy industrialist who left Jo an $80,000 inheritance, which launched him into a successful life in business.

In the early 1850s Shelby attended Transylvania University and started a hemp factory and saw mill in Lafayette County, Missouri. When the mill burned down in 1858 he moved his hemp operation to Waverly, Missouri, and as a responsible citizen he donated funds to start the "Shelby College" near Alma, Missouri.

When pro-union guerillas burned down the home and hemp mill of Mrs. Shelby's parents, Jo became involved on the pro-slavery side in the bloody border wars of western Missouri and eastern Kansas. He was offered a commission as brigadier general in the Union Army but declined believing that Missourians' personal rights were being violated.

Shelby joined the rebel cause in 1861 and raised a company of cavalry that eventually grew to regiment size and became known as the "Iron Brigade." He led his men through some of the most serious fighting in the trans-Mississippi area, eventually being recognized as one of the best cavalry leaders west of the Mississippi.

Still recovering from a serious wound to the arm Shelby was promoted to major general by Kirby Smith on May 10, 1865, several days after Lee's surrender in Virginia. The promotion never went through due to the chaotic situation of the time.

Shelby, however, never doubted where his loyalties lay. The Confederacy had collapsed, and though he tried to revive the battle for Southern independence, he now committed himself to his men. In June 1865 he led some 700 volunteers across Texas to exile in Mexico. Though he would have supported the Juaristas, when his men voted to join Maximilian he stood by them. When they reached Mexico City Shelby offered his men to serve Maximilian as a bulwark against French withdrawal, but the Emperor refused. Shelby then attempted to establish himself as a settler, farmer, and businessman, but came finally to realize that there was no future for him in Mexico and he returned to Missouri in 1867.

Back home in Missouri Shelby resumed his farming and business enterprises and in 1890 was appointed U.S. Marshal for Kansas City. As he had, for better or worse, before he returned home from Mexico, Shelby did what he thought was right come what may. To the end of his life he maintained friendships with his old compatriots from the Kansas/Missouri conflict and the Civil War and associated openly with some of the most notorious of them; men like Jesse and Frank James and Cole Younger. He even appeared as a character witness, along with John Newman Edwards, at the trial of Frank James.

Jo Shelby died February 13, 1897, after a prolonged bout of pneumonia. His funeral was one of the largest in Kansas City history. His grave is in the Forest Hills Cemetery in Kansas City, Missouri.

James Edwin Slaughter

June 12, 1827 – January 2, 1900

James E. Slaughter, a great-nephew of President James Madison, was born in Cedar Mountain, Virginia. He went to Virginia Military Institute but when the War with Mexico began in May 1846, he withdrew from VMI to accept a commission in the United States Army. He served under the command of General Winfield Scott on his march from Vera Cruz to Mexico City. After the war he remained in the Army until after the Civil War began. In May 1862 he resigned from the United States Army and joined the Confederate Army and soon became a captain of artillery.

Soon after his promotion to brigadier general in March, 1862 he was transferred to Fort Brown, Texas, where he became ill with laryngitis and dengue which cost him nearly a year of sick leave for convalescence. Slaughter went on to command Confederate forces in Texas, leading his troops to a Confederate victory at the Battle of Palmito Ranch on May 13, 1865, more than a month after Lee's surrender in Virginia.

Slaughter sought refuge from the Reconstruction in Mexico and stayed there until 1867, when he returned to the United States and worked as a civil engineer in Mobile, Alabama. As the years passed Slaughter's health deteriorated and his doctors encouraged him to seek a warmer climate in hopes of relieving his symptoms. He returned to Mexico but died there of pneumonia on January 2, 1900. He was buried in the American cemetery in Mexico City.

Alonzo William Slayback

July 4, 1838 – October 13, 1882

Colonel Alonzo W. Slayback was born near Lexington, Missouri, and graduated in 1856 from Lexington's Masonic College. He later taught school, studied law in St. Louis, and began his law practice in 1857. On April 14, 1859 he married Alice Amelia Waddell, daughter of William B. Waddell, a founder of the famed Pony Express.

When the Civil War came Slayback joined General Sterling Price's Missouri State Guard. Before the army was to move toward Arkansas he slipped back to Lexington to visit his wife and fell prisoner to Federal troops on the night of July 17, 1862. He spent the next three weeks as one of the first POWs of the Civil War. Around the first of August he escaped after hitting a guard with a bucket of water and fleeing for his life through a fusillade of gunfire and into the woods.

Slayback was promoted to colonel in Price's State Guard and was soon transferred to a cavalry detachment under John S. Marmaduke. After suffering a bout with typhoid fever he was commissioned by Kirby Smith to raise a regiment of cavalry in southeast Missouri with the rank of colonel. His regiment was attached to General Jo Shelby's Iron Brigade for the rest of the war and remained by Shelby's side during their great adventure to Mexico after the war.

On the march to Mexico Shelby's brigade of exiles paused crossing of the Rio Grande at Eagle Pass, Texas, to ceremoniously "bury" their battle flag by weighting it with stones and sinking it in the river. Slayback wrote the well-known poem, "The Burial of Shelby's Flag," commemorating the event.

Slayback remained with Shelby at Cordoba for a while before traveling to Havana and on to New York. Eventually he returned to St. Louis where he resumed his law practice and became an active civic leader. He had lived a life of action and brought home from the war a quick temper and heightened sense of personal honor. He and John Cockerill, editor of the *Post Dispatch* newspaper, had a long-standing personal disagreement, and when on October 13, 1882, Slayback stormed into Cockerill's office demanding an apology concerning an old grudge between them, Cockerill shot Slayback dead.

Cockerill barely escaped prosecution, pleading self-defense because a gun had been found on Slayback's body after the shooting. Witnesses, however, stated that Cockerill shot Slayback as he was removing his coat in preparation for a fight or to slap his antagonist. The bullet from Cockerill's gun struck Slayback in the armpit while his arms were raised to remove his coat and penetrated through the heart killing him almost instantly. Many continued to believe that Cockerill murdered Slayback to avoid a beating. An employee of the *Post Dispatch* confirmed these suspicions when he revealed years later that the pistol found in Slayback's coat was obtained from a policeman and placed there after the murder to support Cockerill's claim of self-defense.

The body of Colonel Alonzo Slayback was buried temporarily in St. Louis, then moved and reinterred next to his father in Lexington's Macpelah Cemetery.

Edmund Kirby Smith

May 16, 1824 – March 28, 1893

Lieutenant General Kirby Smith was born in St. Augustine, Florida where his father, Joseph Lee Smith, held a Federal judgeship. He enrolled in the United States Military Academy in 1841 where his classmates called him "Seminole" because of his Florida origins. He graduated in 1845 and was commissioned a second lieutenant in the 5th U.S. Infantry in spite of his nearsightedness. He wore spectacles for the rest of his life.

During the Mexican-American War he served under Generals Zachary Taylor and Winfield Scott and participated in the Battles of Palo Alto and Resaca de Palma. In the decade and a half after the war Smith's health was under almost constant attack. He suffered two serious bouts with cholera, had yellow fever while serving on the lower Rio Grande, and received a ghastly thigh wound in a fight with Comanche Indians at Old Fort Atchison, Texas.

Smith joined the Confederate Army in 1861 with the rank of brigadier general and was promoted to major general in October. At the Battle of First Manassas he was shot from the saddle when a bullet struck him in the shoulder. He later served directly under General Richard Taylor and led Confederate troops to victory against Union general Banks during the Red River Campaign in

1864. For the last year of the war Smith commanded the Trans-Mississippi Department, which the fall of Vicksburg had left cut off from the Confederacy east of the Mississippi River.

After the war Smith though very ill still left the country for exile in Mexico, but soon returned to his native country to become president of the Atlantic and Pacific Telegraph Company. In 1875 he served for a short time as president of the University of Nashville, leaving to become professor of mathematics at the University of the South in Sewanee, Tennessee.

General Smith's twenty-two year career at the University of the South came to an end with his death on March 28, 1893. He is buried in the University Cemetery.

Walter Husted Stevens

August 24, 1827 – November 12, 1867

Brigadier General Walter Husted Stevens was born in Penn Yan, New York and graduated fourth in his class from West Point in 1848. He joined the Army Corps of Engineers and served throughout Texas and Louisiana for the next thirteen years. Among his assignments was repairing the fortifications defending the approaches to New Orleans and surveying the rivers and harbors of Texas.

Stevens resigned his commission in the U.S. Army and joined the Confederate Army as a Captain of Engineers and served as General Beauregard's chief engineering officer at the Battle of Bull Run. He served in the Peninsula Campaign in 1862 until General Joseph Johnston was wounded at Seven Pines. He was promoted to colonel under the new commander of the Army of Northern Virginia, Robert E. Lee and later became chief engineer of the Army with the rank of brigadier general. During this period his chief duty was the defense of Richmond. When the southern capital fell on April 2, 1865, Stevens was the last uniformed man to abandon the city, but was with Lee's army when he surrendered to Grant at Appomattox.

Like other Confederates he could not accept defeat and went into exile in Mexico in August 1865. There he became Maximilian's construction engineer and superintendent of the Imperial railroad from Vera Cruz to Mexico City. He remained in charge after the fall of Maximilian's government and protected the property of the railroad company through the chaos and fighting that ensued.

Stevens died of yellow fever in Vera Cruz on November 12, 1876. His remains were returned to Virginia and are buried in the Hollywood Cemetery in Richmond.

Alexander Watkins Terrell

November 3, 1827 – September 8, 1912

Brigadier General Alexander W. Terrell was born in Patrick County, Virginia, but in 1831 the family migrated to Boonville, Cooper County, Missouri. He began a law practice in the Booneville office of Judge Peyton R. Hayden upon graduation from the University of Missouri. In 1852 Terrell moved to Austin, Texas and was elected judge of the State's Second District in 1857.

At the outbreak of the Civil War he joined the 1st Texas Cavalry Regiment, Arizona Brigade, rising quickly in rank to lieutenant colonel and was promoted to brigadier general and given command of "Terrell's Texas Cavalry Regiment" by General Kirby Smith.

Terrell fought in a number of engagements culminating in the Red River Campaign in which he led his troops with distinction. When he received the news of Kirby Smith's surrender he fled to Mexico where he served briefly as a battalion commander in Maximilian's Imperial Mexican Army. Terrell's stay in Mexico was short, however, and within a year he returned to Texas to practice law in Houston.

General Terrell lived a long and eventful life following the war and his adventure in Mexico. In 1893 he was named minister to Turkey in Grover Cleveland's administration, but his most lasting accomplishments came as a result of his writing and leadership of the Texas State Historical Society in Austin. He wrote his memoirs, *Recollections and Reminiscences*, a portion of which was published in 1933 under the title, *From Texas to Mexico and the Court of Maximilian in 1865.* Terrell also annotated thirteen volumes of the Texas Supreme Court decisions and published several articles in the *Southwest Historical Quarterly.*

Terrell died in Mineral Wells, Texas on September 8, 1912, while driving his carriage home from a visit to friends in Virginia. His remains rest in the State Cemetery in Austin, and his memory is memorialized by Terrell County, Texas, which is named in his honor.

John George Walker

July 22, 1822 – July 20, 1893

Major General John G. Walker was born in Jefferson City, Missouri and grew to manhood in St. Louis where he graduated in 1844 from the college that became St. Louis's Washington University. When the war with Mexico broke out in 1846 Walker joined the U.S. Mounted Rifles and was later promoted to the rank of captain for his performance at the Battle of San Juan de los Llanos. Wounded at the Battle of Molina del Rey, Walker remained in the Army until April 1861 when he resigned to join the Army of the Confederate States with an appointment to lieutenant colonel of the 8th Texas Cavalry.

In January 1862 he was promoted to brigadier general serving in the Peninsula Campaign and was wounded at Malvern Hill. Serving under Major General James Longstreet, Walker participated in the Battles of South Mountain and Antietam. In November 1862 he was promoted to major general and transferred to the Trans-Mississippi Department where he took command of twelve regiments at Camp Nelson, Arkansas. Walker formed these regiments into a division that, because of their ability to travel fast and strike suddenly, became known at *Walker's Greyhounds*. The

Greyhounds were one of the first Confederate units to face and defeat Black Union troops at the Battle of Milliken's Bend on June 6, 1863. At Pleasant Hill, Louisiana, on April 9, 1864 he was struck in the groin by a bullet and refused to leave the field until fainting and falling from his horse. Walker played a significant role in defeating Nathaniel P. Banks and his *Army of the Gulf* during the Red River Campaign when on April 8, 1864, Walker's Greyhounds contributed to the Confederate victory at the Battle of Mansfield.

At the end of the war he fled to Mexico and lived there for several years before returning to the United States to become U.S. Consul in Bogota, Colombia. He was also given the position of special commissioner to encourage Latin American republics to attend the Latin American Conference in 1889.

Walker suffered a cerebral hemorrhage in Washington, D.C. and lapsed into a four-day coma. He died on July 20, 1893 and is buried in the Stonewall Jackson Cemetery in Winchester, Virginia

APPENDIX A

AN INTERVIEW WITH SHELBY, 1877

"I am no longer a soldier, as you may see, but a farmer and a man of peace."

"But you led an expedition in Mexico once, general."

"Yes, an expedition of a thousand men. It could have been fifty thousand just as well."

"Tell me something about the first expedition then."

"There were several things which led to that. Some have been told and others haven't. Perhaps the time is as good as any to make them known, inasmuch as the attention of the people and the government is diverted somewhat toward Mexican matters. There were a thousand men in my division who did not want to surrender. If there had been but two I would have felt it my soldierly duty to have stood by those two and to have gone with them into the unknown. Then again, I had ideas, or dreams, or ambitions. I saw or imagined I saw an empire beyond the Rio. This river they call the great river.

"Through Gen. Frank P. Blair I had received, long before the killing of Lincoln, some important information. It was to the effect that, in the downfall of the Confederacy and the overthrow of the Confederates in the east, the Confederates of the west would be permitted to march in to Mexico, drive out the French, fraternize with the Mexicans, look around them to see what they could see, occupy and possess lands, keep their eyes fixed steadfastly upon the future, and understand from the beginning that the future would have to take care of itself. In addition, every disbanded Federal soldier in the Trans-Mississippi department, who desired service of the kind I have indicated, would have been permitted to cross over to the Confederates with his arms and ammunition. Fifty thousand of these were eager to enlist in such an expedition. On my march south from San Antonio to Pedras [sic] Negras I received no less than two hundred messages and communications from representative Federal officers, begging me to wait for them beyond the Rio Grande."

"Do you mean to say, General, that President Lincoln was in favor of the movement you have outlined?"

"I do mean to say so most emphatically. I could show nothing official for my assertion, but I had such assurances as satisfied me, and other officers of either army had such assurances as satisfied them. There was empire in it, and a final and practical settlement of this whole Mexican question."

"Why did the scheme fail?"

"I will tell you why. Before marching into the interior of Mexico from Pedras [sic] Negras, a little town on the Rio Grande opposite Eagle Pass, I called my officers and men about me and stated to them briefly the case. Gov. Blesca [sic], the Juarez governor of the state of Coahuila, was in Pedras [sic] Negras. I had sold him cannon, muskets, ammunition, revolvers, sabers – munitions of war which I had brought out of Texas in quantities – and had divided the proceeds per capita among my men. Gov. Biesca offered me the military possession of New Leon and Coahuila, a commission as major general, and absolute authority from Juarez to recruit a corps of fifty thousand Americans. All these things I told my followers. Then I laid a scheme before them and mapped out for the future a programme [sic] which had for a granite basis, as it were, that one irrevocable idea of empire. But to my surprise and almost despair nearly the entire expeditionary force were resolute and aggressive imperialists. I could not move them from the idea of fighting for Maximilian. They hated Juarez, they said, and they hated his cause. Maximilian had been the friend of the South, so had the French, and so had Louis Napoleon. They would not lift a hand against the imperial government. I did not argue with my soldiers. They had been faithful to me beyond everything I had ever known of devotion, and so I said to them, 'You have made your resolve, it seems – so be it. I will go with you to the end of the earth, and if Maximilian wants us we will bring him fifty thousand Americans.' That afternoon late I thanked Gov. Biesca cordially for all he had done and all he wished to do, and marched with my men toward Monterey, where there was a heavy French garrison under Gen. Jeanningros. Jeanningros had heard of the sale of the war munitions to the Mexicans, and he was furious in his wrath and threatenings. We marched, however, to within two miles of Monterey, drew up in line of battle in front of the cuirassiers covering the northern road, dispatched a flag of truce into the town to know whether it was to be peace or war, and Jeanningros made lines promptly, and were known from that on until the evacuation as being in full sympathy and accord with the French."

"And how did Maximilian receive you, General, and what did he say in answer to the proffer you made to him of your services?"

"There is another feature of those plans which were never fulfilled which might be mentioned. Gen. Preston, of Kentucky, was our negotiator, sent forward to represent to Maximilian and Marshal Bazaine the necessity of organizing for immediate service a corps of fifty thousand Americans. Preston talked eloquently and well, but received simply courteous attention for his pains. Neither yes nor no was said to him. Three times he was accorded an audience by the emperor, and three times he came out from it as he went in. Meanwhile we were marching rapidly and fighting our way toward the capital. Guerrillas beset us night and day. There was an ambushment in front and sometimes on both flanks at once. We whipped everything that encountered us from the Salinas River to Queretaro, losing in killed on the rugged march over one hundred of my best and bravest soldiers. I never left a wounded man except in a town where there was a permanent French garrison. Of [those] wounded there were over one hundred more. On reaching the City of Mexico I sought an interview with the emperor at once, and obtained it through Commodore Maury, then Maximilian's commissioner of immigration. It was a brief but emphatic one. The emperor was gracious yet reticent. He asked me what my men and myself desired. I told him service under the empire. He enquired further of the number of men I had with me, the number I could recruit in six months, and the uses that could be made of them after they were recruited. In six months I promised him fifty thousand veterans. As to their uses my answer was about in substance this: 'I speak as a soldier, and I ask your majesty's pardon for so doing; but my men expect me to bring them back word of this interview. Without foreign help you cannot keep your crown. The French will be forced to evacuate Mexico. Mr. Seward has sworn it, and a million of men in arms are anxious to begin today the enforcement of the Monroe doctrine. The Mexicans are not for you. The church party will not fight; the priests – ostensibly your friends – are as enemies working against you secretly; your Austrian and Belgian troops are too few to hold even your capital, and your native regiments are worse than militia. With fifty thousand American soldiers who are devoted to you and who have been schooled in either army to arms, you can hold your own, consolidate your power, develop this empire of yours, and finally get upon the most amicable terms with your naturally selfish, grasping, and unscrupulous neighbors of the United States. Otherwise, if you do not lose your life, you will surely lose your adopted country.' Maximilian listened attentively as I told him this and much more like it, evidently pleased at my frankness and directness. He asked but one other question before the interview terminated."

"What was that question, General?"

"It was this: 'Cannot diplomacy do for me what you propose to do with arms?' I answered: 'It cannot, your majesty,' and I bowed and left him. Afterward, with Gen. Magruder, I called upon Marshal Bazaine and made almost the same kind of speech to him. Bazaine was a thorough soldier. He saw the situation exactly as we saw it. He knew that he would soon have to get out with all of his army. He knew that without foreign aid Maximilian was lost beyond redemption. He knew of what stuff the imperial Mexican soldiers were made. He knew that out of nine millions of people eight millions were for Juarez and the republic was against Maximilian and the empire, and that of the other million, who were imperialists from sentiment and aristocracy, not fifty thousand would take guns in their hands and stand about their king as became men true to honor or brave for the right. But Bazaine was powerless. His business in Mexico, was primarily to collect through Maximilian the debt owed the French by the Mexicans. Afterward he was left there because Napoleon believed he might establish an empire if the Americans kept on cutting one another's throats until there were left but few throats to cut. Napoleon, however, made the fatal mistake of fighting his battles for empire on the Rio Grande instead of on the Potomac. He landed his expeditionary corps at the wrong place. Instead of sending soldiers inland from Vera Cruz, Tampico, and Matamoros, he should have sent them inland from Charleston, Wilmington, and New Orleans. I said this plainly to Bazaine, and Bazaine shrugged his shoulders and bowed his head."

From *Centennial History of Missouri*

APPENDIX B

DECREE TO ENCOURAGE IMMIGRATION

We, Maximilian, emperor of Mexico, in consideration of the sparseness of population in the Mexican territory in proportion to its extent, desiring to give to immigrants all possible security for property and liberty, in order that they may become good Mexicans, sincerely attached to their new country, and having heard the opinion of our board of colonization, do decree as follow:

ARTICLE 1. Mexico is open to immigration from all nations.

ART 2. Immigration agents shall be appointed, who will be paid by the government, and whose duty it will be to protect the arrival of immigrants, and install them on the lands assigned them, and assist them in every possible manner in establishing themselves. These agents will receive the orders of an imperial commissioner of immigration, specially appointed by us, and to whom, through our minister of improvement, (fomento,) all communications relating to immigration shall be addressed.

ART 3. Each immigrant shall receive a duly executed title, incommutable, of his landed estate, and a certificate that it is free of mortgage.

ART 4. Such property shall be free from taxes for the first year, and also from duties on transfers of property, but only on the first sale.

ART 5. The immigrants may be naturalized as soon as they shall have established themselves as settlers.

ART 6. Immigrants who may desire laborers with them, or induce them to come in considerable numbers, of any race whatever, are authorized to do so, but those laborers will be subject to special protective regulations.

ART 7. The effects of immigrants, their working and brood animals, seeds, agricultural implements, machines and working tools, shall enter free of custom-house and transit duties.

ART 8. Immigrants are exempted from military service for five years, but they will form a stationary militia, for the purpose of protecting their property and neighborhoods.

ART 9. Liberty in the exercise of their respective forms of religious worship is secured to immigrants by the organic law of the empire.

ART 10. Each of our ministers is charged with carrying out such parts of this decree as relate to his department.

Given at Chapultepec on the 5th day of September, 1865.

Maximilian.

By the emperor:

The Minister of Improvement, (Fomento)

MANUEL OROZCO Y BERRA

Sub-secretary, (in the absence of the Minister of Improvement)

APPENDIX C

REGULATIONS

Under article 6 of the foregoing decree we ordain as follows:

1. Under the laws of the empire all persons of color are free by the mere act of touching Mexican soil.

2. They may make contracts with the employer who has engaged, or may engage them, by which such employer shall bind himself to feed, clothe, and lodge them, and give them medical attendance, and also pay them a sum of money according to whatever agreements they may enter into with him; moreover, he shall deposit in the savings bank hereinafter mentioned, for the benefit of the laborer, a sum equivalent to one-fourth of his wages. The laborer shall, on his part, obligate himself to his employer to perform the labor for which he was employed, for a term of not less than five, nor more than ten years.

3. The employer shall bind himself to support the children of his laborers. In the event of the father's death, the employer will be regarded as the guardian of the children, and they will remain in his service until they become of age, on the same terms as those agreed to by their father.

4. Each laborer shall receive a book certified by the local authority, in which book a description of his person, the statement of his place of labor, and a certificate of his life and habits will be entered. In case of a change of employer, the consent of the former employer shall be entered in this book.

5. In case of the death of the employer, his heirs, or whoever may acquire his estate, shall be bound to the laborer in the same manner in which such employer was; and the laborer, on his part, shall be bound towards such new proprietor on the same terms as in his former contract.

6. In case of desertion, the laborer when arrested, shall be placed, without pay, upon the public works, until his employer presents himself to claim him.

7. In case of any injustice of the employer towards the laborers, he shall be brought before a magistrate.

8. Special police commissioners will see to the enforcement of these regulations, and officially prosecute all violators thereof.

9. A savings bank will be established by the government for the following objects:

10. The employers shall deposit in said bank, every month, for the benefit of the laborers, a sum equivalent to one-fourth of the wages each is entitled to under his contract of employment.

11. The laborers can deposit, in addition, in the savings bank, in money, such sums as they may desire.

12. These deposits will bear interest at the rate of five per cent per annum.

13. At the end of his engagement, and on presentation of his book, the laborer shall receive the entire amount of his savings.

14. If at the end of his engagement the laborer wishes to leave his money in the savings bank, he can then receive the interest accrued; or, if he wishes to leave this also, it will be added to his capital and also draw interest.

15. In case a laborer should die intestate or without heirs, his property shall pass to the treasury of the government.

Given at Chapultepec on the 5th day of September, 1865.

Maximilian

By the emperor:
The Minister of Improvement (Fomento,)
MANUEL OROZCO Y BERRA
Sub-secretary, (in the absence of the Minister of Improvement.)

APPENDIX D

REGULATIONS AND INSTRUCTIONS IN FURTHERANCE OF THE "DECREE TO ENCOURAGE IMMIGRATION."

ARTICLE 1. The lands offered for colonization are divided into three classes:

ART. 2. First. Those that are of the public domain and have never been reduced to cultivation.

ART. 3. Second. Those that have been more or less improved as haciendas, the right to dispose of which the government has acquired either by purchase or otherwise.

ART. 4. Third. Private lands and haciendas, the owners of which are disposed to offer them to immigrants on liberal terms for colonization. Many haciendas that are or have been under cultivation may be bought on easy terms, and for less than one dollar per acre.

ART. 5. These private haciendas or plantations sometimes embrace several hundred square miles. Smaller sizes often afford lands and room for a settlement of a dozen or more families.

ART.6. Immigrants are advised to establish themselves at first in settlements or communities as well for mutual protection and assistance as for the benefit of churches and schools, the convenience of mills, blacksmith shops, etc., etc.

ART. 7. It is the policy of the government to encourage settlement upon private as well as public lands; and the same rights, privileges, and exemptions are offered to immigrants who may settle upon the former as are granted to those who settle upon the latter.

ART. 8. Lands of class first are offered in alternate sections, as donations to actual settlers, and in quantities varying from 160 acres for a single man, to 640 for the heads of families, according to circumstances to be explained hereafter.

IMMIGRATION

ART. 9. The government not only invites all well-disposed persons to come and assist in the occupation of its vacant places, but his majesty the emperor, touched by the spectacle of good men struggling with adversity in other lands, tenders hospitality and homes especially to these. Moved by the generous impulses of his nature, he offers them material assistance to enable them to reach this bountiful and beautiful land. To those of them who wish to change their skies, make Mexico their home, and identify themselves with the country, a free passage by sea for their families and effects is offered.

ART. 10. Immigrants are therefore divided into two classes, A and B. The former being of those just alluded to, who by misfortune have lost all their substance; and the latter, those who are less straitened in their means.

ART. 11. Not only a free passage by sea is offered to class A, but when they arrive in the country, a traveling allowance of ten cents the league thence to their new homes will be made for each member of their families, counting as members, also, their apprentices.

ART. 12. Lands of class first will be donated to those immigrants by alternate sections, viz: 160 acres to a single man, and 320 to a man with a family, with a pre-emption right to as much more in each case.

ART. 13. Immigrants of class A, who, after arriving in the country, may prefer to settle upon haciendas or other lands, are at liberty to do so; but in that case they may be required to refund, with interest, the money that may have been advanced in assisting them to reach their new domiciles.

ART. 14. Immigrants of class B, who are those that can afford to pay their own expenses, have the whole country before them. They may establish themselves wherever they can find suitable and available lands. If they prefer the unimproved lands of the public domain, they also can have them free, in alternate sections, but only for actual settlement, at the rate of 320 acres for a single man, and 640 to a man with a family, with the pre-emption right to as much more in each case.

ART. 15. These donations of land to persons, whether of class A or B, are made on condition and with the understanding that they shall, in good faith, proceed forthwith to occupy, subdue and cultivate the land so donated.

ART. 16. The lands of the public haciendas will be offered at government prices, and pro rata according to the actual cost of purchase by the government.

ART. 17. As it regards private lands and haciendas, the field of selection is much larger. Immigrants are free to make their own bargains with proprietors, the government waiving its fees on such transactions, as per article four of the decree.

ART. 18. As it regards the public haciendas, a reservation of improved lands will be made from each to serve during the first years as a common, for the free use of the colonists. The size of this common will be determined by the number of families the lands of the hacienda may be sufficient to accommodate. It will be large enough to give them breadstuffs and vegetables at once, and until they can bring their own lands into cultivation. It will be large enough also to afford space for a village, in case the immigrants should find it desirable, as probably at first they will, to establish themselves in villages. No rent will be charged for the first years for the use of this common.

ART. 19. This reserve or common is ultimately intended for educational purposes, and, after the first years, a ground rent of ten per cent upon the value of the land, but not of the improvements, will be required.

AGENCIES

ART. 20. Agents for immigration will be stationed at convenient points abroad, for the purpose of affording information to the emigrant there, as to this country, its lands, the best way of reaching them, and upon all other subjects pertaining thereto.

ART. 21. Persons wishing to emigrate will first apply to the most convenient agent. The applicant must state his occupation, whether agricultural, mercantile, mechanical, or professional. He must also give his age, with the name, sex, and age of each member of his family, including apprentices. If he requires assistance for the journey, he must state his circumstances, and give satisfactory references as to his character and standing in the community.

PERMITS

ART. 22. The permit of the immigration agent is necessary to entitle the immigrant to the privileges of the decree. Unless he bring with him such a permit into the country, he will neither be entitled to lands, to free entry at the custom-house, nor to any other privileges beyond those accorded to mere strangers.

EFFECTS

ART. 23. Immigrants with such permits may bring in, duty free, all their personal and household effects, their live stock, their implements of husbandry, tools and instruments of all kinds used by them in the pursuit of their trade, are, profession, or calling. But they may not bring, without the payment of duty, any merchandise, or thing for sale, exchange, or barter.

ART. 24. Before embarking, if coming by sea, or leaving home, if coming by land, the emigrant should furnish the agent for immigration with a complete list of the persons and effects that he proposes to carry with him. These effects must be properly packed, marked, and numbered; and the contents of each package, box, or case must be duly certified, in order that they may the more readily pass the custom-house and toll-gates.

ART. 25. Arriving in port, or crossing the boundary, an agent will be there, whose duty it is to give the immigrants such assistance, and afford them such further information, as they may require to speed them on their way.

OFFICE OF COLONIZATION, NO. 13 CALLE SAN JUAN LETRAN,

Mexico, November 6, 1865.

MEXICO, September 27, 1865

Approved: **Maximilian**

END NOTES

INTRODUCTION

1. Herbert Ingram Priestly, *The Mexican Nation, A History*, (New York: Macmillan Company, 1938), pp. 345-346
2. Carl E. Bock presents a detailed survey of the agreements in *Prelude to Tragedy, The Negotiations and Breakdown of the Tripartite Convention of London, October 31, 1961* (Philadelphia: University of Pennsylvania Press, 1966).
3. Henry Bamford Parkes, *A History of Mexico* (Boston Press, 1950), pp. 253-255.
4. Ivie E. Cadenhead, Jr., *Benito Juarez* (Twayne Publishers, Inc. 1973), pp.87-88; Charles Allen Smart, *Viva Juarez* (J.B. Lippincott Co. 1963) pp. 237-238
5. Joseph A. Crisp, *Understanding the Second Empire of Mexico and American Involvement in the War for Monarchy* (Designed

by Trevor Stanley for the Austrian Monarchist Alliance). pp. 2-6; Charles Allen Smart, *Viva Juarez* (J.B. Lippincott Co. 1963). pp. 237-238

6. Ibid. *Viva Juarez* pp. 237-238
7. Ibid. Viva Juarez pp. 237-238

CHAPTER ONE

1. James Morton Callahan, *Diplomatic History of the Southern Confederacy* (New York: Frederick Ungar Publishing Co. 1901) p. 72
2. Frank Lawrence Owsley, *King Cotton Diplomacy, Foreign Relations of the Confederate States of America* [Hereinafter referred to as *King Cotton Diplomacy*] (Chicago: University of Chicago Press, 1959) p. 89
3. Ibid., p. 96
4. Robert Lee Kerby, *The Confederate Invasion of New Mexico and Arizona, 1861-1862* (Westernlore Press, 1958) pp. 47-48
5. Ibid., p. 43
6. *Handbook of Texas Online*, s.v., "Vidaurri, Santiago," June 6, 2001
7. Vidaurri led an adventurous life, but his activities along the border eventually caught up with him when he was captured, court-martialed, and shot to death by firing squad in the square of Santo Domingo in June 1867.
8. J. Fred Rippy, "Mexican Projects of the Confederates," *Southwestern Historical Quarterly*, XXXII (April, 1919), p. 293
9. *The War of the Rebellion: A Compilation of the Official Records of the Union and Confederate Armies,* [Hereinafter referred to as the *O.R.*] Ser. I, Vol. IV, Pt. II (Washington: United States Government Printing Office, 1880-1901), pp. 881-882
10. J. Fred Rippy, *The United States and Mexico* (New York: Alfred A. Knopf. 1926), p. 233

11. James Morton Callahan, *American Foreign Policy in Mexico Relations* (New York: Cooper Square Publishers, Inc. 1967) pp. 280-281
12. William T. Windham, "The Problem of Supply in the Trans-Mississippi Confederacy," *Journal of Southern History,* XXVII (May, 1961) p. 162
13. Owsley, *King Cotton Diplomacy,* p. 103
14. Robert W. Delaney, "Matamoros, Port for Texas During the Civil War," *Southwestern Historical Quarterly,* LVII (April, 1955) p. 473
15. O.R. Series I Vol. XLVIII, Part I, p. 1,279
16. Delaney, p. 480
17. O.R. Series I Vol. I XXVI, Part II, pp. 395-396
18. Ibid., pp. 399-401
19. Delaney, p. 479
20. O.R. Series I, Vol. XXVI, Part II, pp. 67-70
21. Jennie Edwards, comp., *John N. Edwards, Biography, Memoirs, Reminiscences and recollections. Also a reprint of Shelby's Expedition to Mexico, An Unwritten Leaf of the War.* [Hereinafter referred to as *Shelby's Expedition to Mexico.*] (Kansas City: Jennie Edwards, Publisher, 1889) p. 261
22. Ibid., pp. 284-285; Robert Lee Kerby, *Kirby Smith's Confederacy: The Trans-Mississippi South, 1863-1865* (New York: Columbia University Press, 1972) pp. 389-390
23. O.R. Series I, Vol. XV, Part I, pp. 996-998
24. Ibid. Series I, Vol. XLVIII, Part I, pp. 512-513
25. Andrew Rolle, *The Lost Cause: The Confederate Exodus to Mexico*, University of Oklahoma Press, 1965) p. 28; O.R. Series I, Vol. XV, Part I, pp. 996-997Rolle, pp. 82-84; John Denny Ripley, *Santos Benavides: His Influence on the Lower Rio Grande, 1823-1891,* (PhD dissertation, Texas Christian University, 1976); Jerry Don Thomas, "A Stand Along the Border: Santos Benavides and the Battle for Laredo," *Civil War Illustrated,* August 1980
26. Jefferson Davis, *The Rise and Fall of the Confederate Government* (New York: D. Appleton and Company, 1881) pp. 612-619

27. The Dahlgren papers did not find their way back to Federal control until after the war. Union Secretary of War Edwin M. Stanton sent for the documents and probably destroyed them, as they have never been seen since. Historians have debated the authenticity of the part of the document that called for the burning of Richmond and the assassination of Davis and his cabinet: "The men must be kept together and well in hand, and once in the city it must be destroyed and Jefferson Davis and Cabinet killed." Since only copies of the documents are extant, the debate goes on with both sides producing convincing arguments for the veracity of their beliefs concerning the matter.
28. Callahan, p. 256-257
29. William C. Harris, *The Hampton Roads Peace Conference: A Final Test of Lincoln's Presidential Leadership,* (Published by the University of Illinois Press on Behalf of the Abraham Lincoln Association) Vol. 21, No. 1, Winter 2000, p. 89
30. Davis to Blair, Jan. 12, 1865, Collected Works, 8:275
31. David Herbert Donald, *Lincoln* (New York: Simon & Schuster, 1995) pp. 557, 559-560
32. Ibid., p. 560
33. James M. McPherson, *Battle Cry of Freedom* (Oxford University Press, 1988) pp. 821-823
34. Lincoln to Stevens, Collected Works, 8:287
35. George G. Meade, *The Life and Letters of George Gordon Meade, Major General United States Army,* 2 Vols. (New York: Charles Scribner's Sons, 1913) Vol. 2, P. 260
36. O.R. Series I, Vol. XLVIII, Part I, pp. 512-513
37. Ibid., p. 512
38. Ibid., Series I, Vol. XV, Part II, pp. 457-458

CHAPTER TWO

1. James D. Richardson, comp. *Messages and Papers of the Confederacy*, Vol. II (Nashville: United States Publishing Company, 1905) p. 49

2. James Morton Callahan, *American Foreign Policy in Mexican Relations,* (New York: Cooper Square Publishers, Inc., 1967) p. 298
3. The Confederate government, nevertheless, did not cease their efforts to win French recognition. (Callahan, *Diplomatic History*, p. 253)
4. O.R., Serial I, Vol. XXVI, part ii, p. 140
5. Ibid., pp. 140-151
6. Ibid., Serial I, Vol. XLVIII, part i, pp.1358-1359
7. Many had already given up the Confederacy as a lost cause, deserted their units and made their way to Mexico. At least one detachment sent to bring back a group of Mexico-bound deserters joined them instead, and there was open talk in Shelby's division of going to Mexico instead of returning to Missouri. (Ibid., Serial I, Vol. XLVIII, part ii, p.45)
8. Ibid., Serial I, Vol. XLVIII, part i, pp. 1292-1293
9. Jennie Edwards, p. 233.
10. Daniel O'Flaherty, *General Jo Shelby, Undefeated Rebel* (Chapel Hill: University of North Carolina Press, 1954), p. 229
11. Jennie Edwards, pp. 233-234
12. Ibid., pp. 238-240
13. Alexander Watkins Terrell, *From Texas to Mexico and the Court of Maximilian in 1865* (Dallas: Book Club of Texas, 1933), pp. 4-5

CHAPTER THREE

1. O'Flaherty, p. 225
2. Edwin C. Bearss, Chief Historian Emeritus, National Park Service. Quote from remarks made in presentation before Scottsdale Civil War Roundtable, Scottsdale, AZ, February 2005.
3. Davison, Eddy W. and Foxx, Daniel, *Nathan Bedford Forrest: In Search of the Enigma*, (Pelican Publishing Co., 2007), p. 501
4. Jennie Edwards, pp. 239-240

5. Ibid.
6. Ibid.
7. John Newman Edwards, pp. 8-9
8. Ibid., pp. 10-11
9. Ibid., p. 11
10. Ibid., p. 13
11. Ibid., pp. 13-14
12. Ibid., p. 14
13. Ibid., pp. 14-15
14. Jenny Edwards, pp. 245-247
15. Ibid., pp. 248-249
16. John Newman Edwards, pp. 16-17
17. O'Flaherty, p. 240
18. John Newman Edwards, p. 18
19. Ibid., pp. 18-19
20. Carl Coke Rister, "Carlotta, A Confederate Colony in Mexico," *Journal of Southern History,* XI (February, 1945), p. 36. [A list of "new arrivals" was also printed in the *Mexican Times*, September 23, 1865, which included these names and many more.]
21. Terrell, p. 25
22. John Newman Edwards, p. 21
23. Ibid., p. 22
24. Jennie Edwards, p. 261
25. Stevens, *Centennial History of Missouri,* pp. 297-298 [See Appendix A]
26. Ibid., pp. 298-299
27. John Newman Edwards, p. 24
28. Ibid., pp. 24-25

CHAPTER FOUR

1. Rolle, p. 59
2. Edwin Adams Davis, *Fallen Guidon: The Saga of Confederate General Jo Shelby's March to Mexico.* (College Station, Texas: Texas A&M Press, 1999), p. 77; O'Flaherty, p. 250

3. O'Flaherty, pp. 254-257;
4. John Newman Edwards, pp. 29-31
5. Davis, p. 92
6. Edwards, pp. 44-45
7. Ibid., pp. 44-45
8. Ibid., p. 41; Jennie Edwards, p. 288
9. O'Flaherty, pp. 270-272; John Newman Edwards, p. 47-49
10. Garcia, "Documentos Para la Historia de Mejico," XXXIII, 245-249 (quoted in J. Fred Rippy, "Mexican Projects of the Confederates," *Southwestern Historical Quarterly Online*, Vol. 22, No. 4, updated April 9, 2004.)
11. Edwards, p. 55
12. Ibid., pp. 56-57
13. Ibid., p. 67
14. Ibid., p. 74
15. Davis, p. 149
16. John Newman Edwards, pp. 90-91
17. Ibid., pp. 91-91
18. Ibid., p. 92
19. Ibid., p. 92
20. Ibid., pp. 92-94
21. Ibid., pp. 94-95
22. Ibid., p. 97
23. Ibid., p. 97-98

CHAPTER FIVE

1. O.R., Series I, Vol. XLVIII, Pt ii, p. 1015
2. Ibid., pp. 923-924. Phillip H. Sheridan, *Personal Memoirs of P. H. Sheridan*, Vol. II (New York: Charles L. Webster & Company, 1888), p. 213
3. O.R. Series I, Vol. XLVIII, pt. ii, p. 1148
4. Ibid., p. 1148

5. Jennie Edwards, *Shelby's Expedition to Mexico*, pp. 360-362; Centennial History of Missouri
6. Alfred J. and Kathryn Abbey Hanna, "The Immigration Movement of the Intervention and Empire as Seen Through the Mexican Press," *Hispanic American Historical Review.* XXVIII (May, 1947), p. 220
7. Count Egon Caesar Corti, *Maximilian and Charlotte of Mexico, trans.* By Catherine Alison Phillips. Vol. I (New York and London; Alfred A. Knopf, 1929), p. 310
8. Hanna, "The Immigration Movement of the Intervention." pp. 117-118
9. It was only natural that many Mexicans should entertain misgivings about Southern immigration to their country. There was a general feeling that it was the South with its desire for more slave territory that had instigated the War between the United States and Mexico and other troubles in the 1840s and 1850s.
10. Report of the Junta of Colonization, quoted in Hanna, "The Immigration Movement of the Intervention." p. 239
11. W. C. Nunn, *Escape from Reconstruction* (Fort Worth: Texas Christian University, 1956, pp. 27-28
12. Sara Yorke Stevenson, *Maximilian in Mexico, A Woman's Reminiscences of the French Intervention, 1862-1867* (New York: The Century Co., 1899), pp. 178-179
13. *United States Senate Executive Documents, 39th Cong., 1st sess.*, No. 8, 1865-1866, pp. 17-20; 24-25
14. Nunn, p. 28
15. Charles Lee Lewis, *Matthew Fontaine Maury, The Pathfinder of the Seas* (Annapolis: The United States Naval Institute, 1927), p. 186-189
16. H. Montgomery Hyde, *Mexican Empire, The History of Maximilian and Carlota of Mexico.* (London: MacMillan & Co. Ltd, 1946), pp.193-194
17. Diana Fountain [sic] Maury Corbin, *A Life of Matthew Fontaine Maury* (London: Sampson Low, Marston, Searle, & Rivington, Limited, 1888), p. 230
18. Ibid., pp. 230-232
19. Lewis, pp. 190-191

20. *United States House of Representatives Executive Documents, Diplomatic Correspondence, 39th Congress., 2nd sess., No. 1, part 3, 1866,* p. 206 (See Appendix B.)
21. Slavery was constitutionally banned in Mexico. The term "apprentice," therefore typically referred to former slaves who might accompany white southerners to the colonies.
22. *United States House of Representatives Executive Documents, Diplomatic Correspondence, 39th Congress., 2nd sess., No. 2, part 3, 1866*, p. 207-209
23. Corbin, pp. 234-235
24. Ibid., pp. 230-232
25. Ibid., p. 232
26. *Mexican Times*, June 16, 1866, p. 2
27. Ibid., pp. 232-232
28. *House Executive Documents, Diplomatic Correspondence, 39th Congress. sess., No. 1, Part 3, 1866,* p. 206
29. Ibid., pp. 206-207
30. Ibid., p. 212
31. Corbin, pp. 242-243
32. Allen received a subsidy of $10,000 from the Imperial Government. (*Mexican Times*, Jan. 22, 1867, p. 2)
33. Ibid., Sept. 16, 1865
34. Corbin, pp. 247-248
35. *Mexican Times*, Oct. 14, 1865, p. 2
36. Hill, p. 110
37. The prize fugitive, President Jefferson Davis, was captured on May 10 in Irwinville, Georgia, and was imprisoned in Fort Monroe, Florida for two years.
38. James D. Bulloch, *The Secret Service of the Confederate States in Europe*, 2 Vols. (New York, London: Thomas Yoseloff, 1959), p. 415
39. Edgefield Advertiser, Oct. 11, 1865, quoted in Hill, *The Confederate Exodus to Latin America*, pp. 110-111. (The Southern Colonization Society was organized to establish a Confederate refuge in Brazil.)
40. *Mexican Times*, Feb. 10, 1866, p. 3

41. William D. Hoyt, Jr. "New Light on General Jubal Early After Appomattox," *Journal of Southern History*, XI (Feb. 1943), p. 116
42. Mexican Times, Nov. 11, 1865, p. 2

CHAPTER SIX

1. *House Executive Documents, Diplomatic Correspondence, 39th Cong., 2nd sess., No. 1, Pt. 3, 1866*, p. 209-212
2. The prospectus of the American Mexican Emigrant Company was printed in *Mexican Times*, Dec. 2, 1865
3. Luther E. Chandler, "The Career of Henry Watkins Allen" (unpublished Ph.D. dissertation, Louisiana State University, 1940), pp. 260-261
4. Russell, Thomas J., p. 92
5. Rister, "Carlota, A Confederate Colony in Mexico," p. 41
6. *Senate Executive Documents, Diplomatic Correspondence, 40 Cong., 2 sess., No. 2, 1867-1868*, p. 491; *Mexican Times*, Dec. 23, 1865, p. 3
7. Ibid., Oct. 7, 1865, p. 3
8. Ramon Eduardo Ruiz, ed., *An American in Maximilian's Mexico 1865-1866; The Diaries of William Marshall Anderson* (San Marino, Calif.; The Huntington Library, 1959), p. 47
9. *Mexican Times*, Oct. 14, 1865
10. Hill, *The Confederate Exodus to Latin America*, pp. 316-317
11. *New York Herald*, quoted in *Mexican Times*, Nov. 18, 1865
12. *Mexican Times*, Nov. 18, 1865
13. Ibid., Dec. 9, 1865
14. Hill, p. 114
15. Corbin, pp. 238-239
16. Hill, p. 316
17. Chandler, p. 262
18. *The Matamoros Ranchero* reported "not a day passes that exiles from the old Confederate states do not enter the Empire from this point." (Reprinted in *Mexican Times*, Oct. 21, 1865)
19. *Mexican Times*, June 16, 1866

20. *Papers Relating to Foreign Affairs, Annual Message of the President, 39 Cong., 2 sess., Pt. 3, 1866,* p. 216
21. *Mexican Times*, Oct. 29, 1866
22. Hill, p. 315
23. *Mexican Times*, Oct. 7, 1865
24. Chandler, pp. 270-271
25. Ibid., p. 271
26. *Senate Executive Documents, Diplomatic Correspondence, 40 Cong., 2 sess., No. 2, 1867-1868*, p. 492; Chandler, p. 271
27. *Mexican Times*, Jan. 27, 1866
28. Ibid., Feb. 3, 1866

CHAPTER SEVEN

1. Ralph Roeder, *Juarez and His Mexico*, Vol. II (New York: Viking press, 1947), pp. 598-599
2. Jose Luis Blasio, *Maximilian, Emperor of Mexico: Memoirs of his Private Secretary*, Translated by Robert Hammond (New Haven: Yale University Press, 1934), pp. 60-63
3. The Convention of Miramar was signed on the day Maximilian accepted the Mexican throne, April 10, 1864, at the Prince's summer palace on the Adriatic. It was agreed that French forces should be reduced to 25,000 in 1866, to 20,000 in 1867, and that Maximilian's Mexican forces should replace all the French troops within six years. (Priestly, *The Mexican Nation*, p. 355)
4. Jennie Edwards, *Shelby's Expedition to Mexico,* pp. 360-361
5. *Mexican Times*, Nov. 4 and 25, 1866
6. Ruiz, *The Diaries of William Marshall Anderson*, p. 117
7. *Mexican Times*, Feb. 3, 1866
8. Jennie Edwards, *Shelby's Expedition to Mexico*, p. 405
9. *Daily Picayune* (New Orleans), Mar. 1, 1866
10. Ruiz, p. 35
11. *Mexican Times,* June 16, 1866
12. Ibid., Nov. 11, 1865
13. Ruiz, p. 35

14. *Mexican Times*, Feb. 17, 1866
15. Roeder, p. 599
16. Parkes, *A History of Mexico*, p. 267
17. *Mexican Times*, March 24, 1866
18. *Senate Executive Documents, Diplomatic Correspondence, 40 Cong., 2 sess., No. 2 (1867-1868)*, p. 488
19. Ibid., p. 494
20. Thomas J. Russell, "Adventures of a Cordoba Colonist," *Southern Magazine,* I and II (July, August, 1872), pp. 90-98
21. Ibid., p. 157; *Mexican Times*, June 16, 1866
22. Russell, p. 94
23. Ibid., pp. 94-97
24. Ibid., pp. 99-100
25. Ibid., p. 159
26. Ibid., 90-98; *Mexican Times*, June 16, 1866
27. *Mexican Times*, June 23, 1866
28. *Mexican Times*, June 16, 1866
29. Ibid.
30. Ibid., June 23, 1866
31. "The Cordoba Colony," *Kansas Magazine*, I (1872-1873), p. 496
32. John Samuel Ezell, *The South Since 1865* (New York: MacMillian Co., 1963), p. 36
33. *Mexican Times*, June 16, 1866

Daily Picayune (New Orleans), July 4, 1866

Ruiz, The Diaries of William Marshall Anderson, pp. 114-115

34. Mexican Times, Jan. 13, 1866
35. Chandler, "The Career of Henry Watkins Allen," p. 239
36. *Daily Picayune* (New Orleans), July 4, 1866
37. *New York Tribune*, June 22, 1866, p. 17
38. *Senate Executive Documents, 40 Cong., 1st sess., & Special Sess., No. 22, 1867*, p. 293
39. Parkes, *A History of Mexico*, pp. 269-271
40. *House Executive Documents, Diplomatic Correspondence, 40th Cong., 1st sess., No. 13, 1867-1868*, pp. 352-353
41. *Mexican Times*, Jan. 15, 1867
42. Ibid., Feb. 23, 1867
43. Ibid., June 17, 1867

CHAPTER EIGHT

1. Mexican Times, *Oct. 8, 1866*
2. *Ibid., Nov. 19, 1866*
3. *John Newman Edwards, p. 117*
4. *O'Flaherty, p. 311*
5. *J. O. Shelby to R. J. Lawrence, Feb. 2, 1866. Printed in* Missouri Historical Bulletin, *Vol. IX (Jan 1953), p. 217*
6. *O'Flaherty, p. 311*
7. *Ibid., p. 331-333*

CHAPTER NINE

1. *Daily Picayune* (New Orleans), Feb. 3, 1867
2. Hanna, "The Immigration Movement of the Intervention," p. 220; *Mexican Times*, June 23, 1866

House Executive Documents, Diplomatic Correspondence, 40th Cong., 1st sess., No. 13, 1867-1868, p. 366

3. *Daily Picayune* (New Orleans), Feb. 3, 1867
4. "The Cordoba Colony," p. 496
5. *House Executive Documents, Diplomatic Correspondence, 40th Cong., 2nd sess., No. 13, 1867-1868*, p. 366
6. *Daily Picayune* (New Orleans), Feb. 3, 1867
7. *Matamoros Ranchero*, July 20, 1867, quoted in Nunn, *Escape from Reconstruction*, p. 112
8. Mexican Dispatches, XXXVII, quoted in Hill, *The Confederate Exodus to Latin America*, p. 322
9. Chandler, *The Career of Henry Watkins Allen*, pp. 270-271
10. A large number of issues of *The Two Republics* were missing in the collections consulted by the authors. With the next available issue, dated October 4, 1883, George W. Clarke is no longer mentioned as editor of *The Two Republics*. J. Mastella Clarke, who had been named in previous issues as assistant

editor, was now given as editor and publisher along with Walter M. O'Dwyer.

11. Frank A. Knapp, Jr. "A New Source on the Confederate Exodus to Mexico: The Two Republics," *Journal of Southern History*, XIX (August, 1953), p. 369

12. *The Two Republics*, July 27, 1867

13. Ibid., Aug. 4, 1867

14. Ibid., Sept. 28, 1867

15. *House Executive Documents, Diplomatic Correspondence, 40th Cong., 2nd sess., No. 13, 1867-1868*, p. 366

16. This statement was obviously made to the best of Clarke's knowledge. (*The Two Republics*, Oct. 5, 1867.)

17. Ibid., Nov. 20, 1867

18. Ibid., Feb. 19, 1868

19. Ibid., Nov. 27, 1867

20. Ibid., Dec. 14 and 28, 1867

21. Gideon Lincecum, *Autobiography* (M.S. Archives Collection, University of Texas Library, Austin), quoted in Nunn, *Escape From Reconstruction,* pp. 74-75

22. *The Two Republics*, July 27, 1867

23. Ibid., Nov. 16, 1867

24. Ibid., Jan. 29, and Mar. 18, 1868

25. Mexican Dispatches, XXXVII, quoted in Hill, *The Confederate Exodus to Latin America*, pp. 324-325

26. Ibid., pp. 325-326

27. *The Two Republics*, Jan. 12, 1876

28. Ibid., Feb. 2, 1878

29. Ibid., Jan. 27, 1876

CHAPTER TEN

1. *Atlanta Courier*, reprinted in *The Two Republics*, May 10, 1876

BIBLIOGRAPHY

BOOKS

Anna, Timothy E. *The Mexican Empire of Iturbide*. Lincoln: University of Nebraska Press, 1990.

Bailey, Thomas A. *A Diplomatic History of the American People*. New York: F.S. Crofts & Co., 1940.

Beezley, William H. *Mexico in World History*. U.S.A., New York: Oxford University Press, 2011.

Blasio, Jose Luis. *Maximilian, Emperor of Mexico: Memoirs of his Private Secretary*. Translated by Robert Hammond Murray. New Haven: Yale University Press, 1934.

Bock, Carl H. *Prelude to Tragedy, The Negotiations and Breakdown of the Tripartite Convention of London, October 31, 1861*. Philadelphia: University of Pennsylvania Press, 1966.

Bulloch, James D. *The Secret Service of the Confederate States in Europe*. 2 Vols.; New York, London: Thomas Joseloff, 1959.

Burke, Ulick Ralph. *A Life of Benito Juarez, Constitutional President of Mexico*. Kessinger Publishing, LLC, 2007.

Cadenhead, Ivie E. Jr. *Benito Juarez*. Twayne Publishers, Inc., 1973.

Callahan, James Morton. *American Foreign Policy in Mexican Relations*. New York: Cooper Square Publishers, Inc., 1967.

______. *Diplomatic History of the Southern Confederacy*. New York: Frederick Ungar Publishing Co., 1901.

Casdorph, Paul D. *Prince John Magruder: His Life and Campaigns*. Wiley, Inc. 1996.

Castel, Albert. *General Sterling Price and the Civil War in the West*. Baton Rouge: Louisiana State University Press, 1993.

Connelley, William Elsey. *Quantrill and the Border Wars.* Cedar Rapids, Iowa: Torch Press, 1910.

Corbin, Diana Fontain [sic] Maury. *A Life of Matthew Fontaine Maury.* London: Sampson Low, Marston, Searle, & Rivington, Limited, 1888.

Corti, Count Egon Caesar. *Maximilian and Charlotte of Mexico.* Translated by Catherine Alison Phillips. 2 Vols.; New York & London: Alfred A. Knopf, 1929.

Crisp, Joseph H. *Understanding the Second Empire of Mexico and American Involvement in the War for Monarchy* (Designed by Trevor Stanley for the Austrian Monarchist Alliance).

Daddysman, James W. *The Matamoros Trade: Confederate Commerce, Diplomacy, and Intrigue*. Newark: University of Delaware Press, 1972.

Davis, Edwin Adams. *Fallen Guidon: The Saga of Confederate General Jo Shelby's March to Mexico.* College Station: Texas A&M University Press, 1995.

Davis, Jefferson. *The Rise and Fall of the Confederate Government.* 2 Vols.; New York: D. Appleton Co., 1881.

Davison, Eddy W. and Foxx, Daniel. *Nathan Bedford Forrest: In Search of the Enigma.* Gretna, Louisiana: Pelican Publishing Co., 2007.

Dawson, Daniel. *The Mexican Adventure*. London: G. Bell and Sons Ltd., 1935.

Donald, David Herbert. *Lincoln.* New York: Simon & Schuster, 1995.

Dunbar, Rowland, ed. *Jefferson Davis, Constitutionalist: His Letters, Papers and Speeches*. 10 Vol. Jackson, Mississippi, 1923.

Edwards, Jennie, comp., *John N. Edwards, Biography, Memoirs, Reminiscences and Recollections.* Also a reprint of *Shelby's Expedition to Mexico, An Unwritten Leaf of the War.* Kansas City: Jennie Edwards, Publisher, 1889.

Edwards, John Newman, *Shelby and his Men: or The War in the West.* Cincinnati: Miami Printing and Publishing Co., 1867.

_____. *Shelby's Expedition to Mexico, An Unwritten Leaf of the War*. Austin, Texas: The Steck Company Publishers, 1872.

Ezell, John Samuel. *The South Since 1865*. New York: Macmillan Co., 1963.

Foster, Gaines M. *Ghosts of the Confederacy*. Oxford University Press, 1987.

Grant, Ulysses S. *Personal Memoirs of U.S. Grant*. New York: Charles L. Webster and Co., 1886

Hall, Frederic. *Invasion of Mexico by the French; and the Reign of Maximilian I*. New York: James Miller, Publisher, 1868.

Hammett, Brian R. *A Concise History of Mexico*. Cambridge University Press, 2006.

Handbook of Texas Online, s.v., "Vidaurri, Santiago," June 6, 2001

Hanna, Alfred J. Hanna, Alfred J. *Flight into Oblivion*. 2nd ed. Bloomington: Indiana University Press, 1959.

Haslip, Joan. *The Crown of Mexico: Maximilian and His Empress Carlota*. New York: Holt, Rinehart and Winston, 1971.

Hearn, Chester G. *Tracks in the Sea: Matthew Fontaine Maury and the Mapping of the Oceans*. New York: McGraw Hill, 2002.

Hyde, H. Montgomery. *Mexican Empire, The History of Maximilian and Carlota of Mexico*. London: Macmillan & Co. Ltd, 1946.

Kerby, Robert. *Kirby Smith's Confederacy: The Transmississippi South 1863-1865*. Tuscaloosa: University of Alabama Press, 1997.

Kerby, Robert Lee. *The Confederate Invasion of New Mexico and Arizona, 1861-1862*. Westernlore Press, 1958

Lewis, Charles Lee. *Matthew Fontaine Maury, The Pathfinder of the Seas*. Annapolis: United States Naval Institute, 1927.

Lincoln, Abraham. *The Writings of Abraham Lincoln*. 8 Vols. Edited by Arthur Brooks Lapsley, New York; G. P. Putnam's Sons, 1906.

Martin, Percy F. *Maximilian in Mexico, The Story of the French Intervention (1861-1867)*. London: Constable and Company Ltd, 1914.

McAllen, M.M. *Maximilian and Carlota: Europe's Last Empire in Mexico.* San Antonio: Trinity University Press, 2014.

McPherson, James M. *Battle Cry of Freedom.* Oxford University Press, 1988.

Meade, George G. *The Life and Letters of George Gordon Meade, Major General United States Army.* 2 Vols.; New York: Charles Scribner's Sons, 1913.

Myer, Michael C. and Beezley, William H. *The Oxford History of Mexico.* Oxford University Press, USA, 2000

Neal, Diane and Kremm, Thomas W. *Lion of the South: General Thomas C. Hindman.* Macon: Mercer University Press, 1993.

Nevins, Allan, ed., and Richardson, James D., comp., *The Messages and Papers of Jefferson Davis and the Confederacy.* 2 Vols.; New York: Chelsea House-Robert Hector Publishers in Association with R.R. Bowker, 1966.

Niles, Blair. *Passengers to Mexico, The Last Invasion of the Americas.* New York and Toronto: Farrar & Rinehart, Inc., 1943.

Nunn, W.C. *Escape From Reconstruction.* Fort Worth: Texas Christian University, 1956.

O'Flaherty, Daniel. *General Jo Shelby, Undefeated Rebel.* Chapel Hill: University of North Carolina Press, 1954.

Owsley, Frank Lawrence. *King Cotton Diplomacy, Foreign Relations of the Confederate States of America.* Chicago: University of Chicago, 1931.

Parkes, Henry Bamford. *A History of Mexico.* Boston: Houghton Mifflin Company, 1950; Cambridge: Riverside Press, 1950.

Parks, Joseph Howard. *General Edmund Kirby Smith, C.S.A.* Baton Rouge: Louisiana State University Press, 1954.

Patterson, Gerard A. *From Blue to Gray: The Life of Confederate General Cadmus M. Wilcox.* Mechanicsburg, PA: Stackpole Books, 2001.

Priestley, Herbert Ingram. *The Mexican Nation, A History.* New York: Macmillan Company, 1938.

Reid, Whitelaw. *After the War; A Southern Tour*. Cincinnati & New York: Moore, Wilstach & Baldwin, 1866.

Richardson, James D., comp., *Messages and Papers of the Confederacy*. 2 Vol.; Nashville: United States Publishing Company, 1905.

Ridley, Jasper. *Maximilian and Juarez*. Phoenix Press, 2001

Rippy, J. Fred. *The United States and Mexico*. New York: Alfred A. Knopf, 1926.

Robertson, William Spence. *Iturbide of Mexico*. New York: Greenwood Press, 1968.

Roeder, Ralph. *Juarez and his Mexico*. 2 Vols.; New York: Viking Press, 1947.

Rolle, Andrew. *The Lost Cause: The Confederate Exodus to Mexico*. Norman: University of Oklahoma Press. 1965

Ruiz, Ramon Eduardo, ed. *An American in Maximilian's Mexico, 1865-1866: The Diaries of William Marshall Anderson*. San Marino, Calif.: Huntington Library, 1959.

Sandburg, Carl. *Abraham Lincoln, the War Years*. New York: Harcourt, Brace & Co., 1936

Sears, Stephan W. *Controversies and Commanders: Dispatches from the Army of the Potomoc*. Boston: Mariner Books, 2000.

Schultz, Duane P. *The Dahlgren Affair: Terror and Conspiracy in the Civil War*. New York: W. W. Norton & Co., 1999.

Sellmeyer, Deryl P. *Jo Shelby's Iron Brigade*. Gretna, LA: Pelican Publishing, 2007.

Settles, Thomas Michael and Campbell, Kimberly. *John Bankhead Magruder: A Military Reappraisal* (Southern Biography). Louisiana State University Press, 2001.

Sheridan, Philip H. *Personal Memoirs of P.H. Sheridan*. 2 Vols.; New York: Charles L. Webster & Company, 1888.

Stein, R. Conrad. *Benito Juarez and the French Intervention (Story of Mexico)*. Morgan Reynolds Publishing, 2007.

Stevens, Walter B. *Centennial History of Missouri*. 6 Vols.; St. Louis, Chicago: S.J. Clarke Publishing Company, 1921.

Stevenson, Sara Yorke. *Maximilian in Mexico; A Woman's Reminiscences of the French Intervention, 1862-1867*. New York: Century Co., 1899.

Smart, Charles Allen. *Viva Juarez*. New York: J.B. Lippincott Co., 1963.

Terrell, Alexander Watkins. *From Texas to Mexico and the Court of Maximilian in 1865*. Dallas: Book Club of Texas, 1933.

Thomas, Sarah A. *Recollections of Henry Watkins Allen*. Thomas Gale, 1969.

Welsh, Jack D. *Medical Histories of Confederate Generals*. Kent, Ohio: Kent State University Press, 1999.

_____. *Medical Histories of Union Generals*. Kent, Ohio: Kent State University Press, 2005.

Williams, Francis Leigh. *Matthew Fontaine Maury: Scientist of the Sea*. New Brunswick, NJ: Rutgers University Press, 1963.

ARTICLES

"The Cordoba Colony." *Kansas Magazine*, I 1872-1873), 492-505.

Box, Sam. "End of the War—Exiles in Mexico," *Confederate Veteran*, Vol. II (March, 1903), 121-123.

Crisp, Joseph A. "Understanding the Second Empire of Mexico and American Involvement in the War for Monarchy," (Designed by Trevor Stanley for the Australian Monarchist Alliance)

Delaney, Robert W. "Matamoros, Port for Texas during the Civil War." *Southwestern Historical Quarterly*, LVII (April, 1955), 473-487.

Diamond, William. "Imports of the Confederate Government from Europe and Mexico." *Journal of Southern History*. VI (November, 1940), 470-503.

Ellison, Simon J. "An Anglo-American Plan for the Colonization of Mexico," *Southwestern Social Science Quarterly*, Vol. XVI (September, 1935), 42-52.

Hanna, Alfred J. "A Confederate newspaper in Mexico." *Journal of Southern* History. XII (February, 1946), 67-83.

_____, "The Role of Matthew Fontaine Maury in the Mexican Empire," *Virginia Magazine of History and Biography*. Vol. LV (April, 1947), 105-125.

______, and Hanna, Kathryn Abbey. "The Immigration Movement of the Intervention and Empire as Seen Through the Mexican Press." *Hispanic American Historical Review*, XXVII (May, 1947), 220-246.

Hanna, Kathryn Abbey. "Incidents of the Confederate Blockade," *Journal of Southern History*, Vol. XI (May, 1945), 214-219.

_____, "The Roles of the South in the French Intervention in Mexico," *Journal of Southern History*, Vol. XX (February, 1944), 3-21.

Harmon, George D. "Confederate Migration to Mexico." *Hispanic American Historical Review*, XVII (November, 1937), 458-487.

Harris, William C. "The Hampton Roads Peace Conference: A Final Test of Lincoln's Presidential Leadership," (Published by the University of Illinois Press on Behalf of the Abraham Lincoln Association), Vol. 21, No. 1, Winter 2000.

Hill, Lawrence F. "The Confederate Exodus to Latin America." *Southwestern Historical Quarterly*, XXXIX (October, 1935; January, April 1936), 100-134, 161-199, 309-326.

_____, "Confederate Exodus to Latin America", Volume 39, Number 3, *Southwestern Historical Quarterly Online*. http//www.1sha.utexas. http//www.lsha.utexas.edu/publications/journals/shq/online /v039/n3/contrib._DIVL2299, html[accessed Thu Apr 10 12:59 CDT 2008]

_____, "Confederate Exodus to Latin America", Volume 39, Number , 4 http//ww.lsha.utexas.edu/publications/journals/ shq/online/v)039/n4 contrib._DIVL4012,html [accessed Thu Apr 10 12:59 CDT 2008]

Hoyt, William D. "New Light on General Jubal A. Early After Appomattox." *Journal of Southern History*, IX (February, 1943), 113-117.

Knapp, Frank A., Jr. "A New Source on the Confederate Exodus to Mexico: The Two Republics." *Journal of Southern History*, XIX (August, 1953), 364-373.

Rippy, J. Fred. "Mexican Projects of the Confederates." *Southwestern Historical Quarterly, XXII (April, 1919), 291-317*

Rister, Carl Coke. "Carlotta, A Confederate Colony in Mexico." *Journal of Southern History*, XI (February, 1945), 33-50.

Russell, Tom J. "Adventures of a Cordoba Colonist." Sou*thern Magazine*, I and II (July, August, 1872), 90-102, 155-166.

Sutherland, Donald E. "Exiles, Emigrants, and Sojourners: The Post-Civil War Confederate Exodus in Perspective," *Civil War History*, XXXI (September, 1985), 237-256.

Thomas, Jerry Don. "A Stand Along the Border: Santos Benavides and the Battle for Laredo," *Civil War Illustrated*, August 1980.

Venter, Bruce. "The Kilpatrick-Dahlgren Raid on Richmond, February 28-March 4, 1964." *Blue & Gray*, Vol. 20, #3.

Windham, William T. "The Problem of Supply in the Trans-Mississippi Confederacy." *Journal of Southern History*, XXVII (May, 1961), 149-168.

GOVERNMENT PUBLICATIONS

Papers Relating to Foreign Affairs, Annual Message of the President, 39th Congress, 2nd Session, Pt. 3, (1866-1867).

United States House of Representatives Executive Documents, 39th Congress, 1st Session, No. 1, Pt. 3 (1866-1867).

United States House of Representatives Executive Documents, 39th congress, 2nd Session, No. 1, Pt. 3 (1866-1867).

United States House of Representatives Executive Documents, Diplomatic Correspondence, 39th Congress, 2nd Session, No. 1, Pt. 3 (1866-1867).

United States House of Representatives Executive Documents, Diplomatic Correspondence, 40th Congress, 2nd Session, No. 13, (1867-1868).

United States House of Representatives Executive Documents, 41st Congress, 3rd Session, No. 1, Pt. 1 (1868-1869).

United States Senate Executive Documents, 39th Congress, 1st Session, No. 8, (1866-1867).

United States Senate Executive Documents, 40th Congress, 1st Session, and Special Session, No. 20, (1867).

United States Senate Executive Documents, Diplomatic Correspondence, 40th Congress, 2nd Session, No. 2 (1867-1868).

The War of the Rebellion: A Compilation of the Official Records of the Union and Confederate Armies. (129 vols.; Washington: United States Printing Office, 1880-1901).

NEWSPAPERS

Daily Picayune (New Orleans), March 1, 1866 – February 12, 1868.

Edgefield Advertiser (Edgefield) October 11, 1865

New York Herald (New York) April 19, 1866

The *Mexican Times* (Mexico City), September 16, 1865 – June 21, 1867.

The Two Republics (Mexico City), July 27, 1867 – January 2, 1895.

UNPUBLISHED MATERIALS

Chandler, Luther E. "The Career of Henry Watkins Allen" (unpublished Ph. D. dissertation, Louisiana State University, 1940).

Maury Papers, Virginia Military Institute archives (available on the Internet).

Ripley, John Denny. "Santos Benavides: His Influence on the Lower Rio Grande" (unpublished Ph. D. dissertation, Texas Christian University, 1976).

University of Missouri Library. Joint Collection: Western Historical Manuscript Collection and State Historical Society Manuscripts, Thomas Westlake, personal memoirs

INDEX

Other Books:

By Daniel Foxx

I Only Laugh When It Hurts

Four Stories of Christmas

With Mary-Helen Foxx

Charlie's Girl

Nellie's Girl

With Eddy W. Davison

Nathan Bedford Forrest: In Search of the Enigma

Rebel Refugees: The Confederate Colonization Movement in Mexico

All books can be purchased at Amazon.com. Authors can be found on Facebook.

www.daniel-foxx.com

ABOUT THE AUTHORS

Daniel Foxx is Professor Emeritus of History at Ottawa University in Phoenix. His other works include a biography of Gen. Nathan Bedford Forrest, with Eddy W. Davison, and two novels with his wife, Mary-Helen. His most recent book is *Four Stories of Christmas.* He lives in Arizona with his family nearby.

Eddy W. Davison writes and speaks on Civil War Topics. He teaches at the Institute of the Americas and is a past member of the board of directors of the Scottsdale Civil War Roundtable. His work has appeared in historical journals and he is the coauthor of *Nathan Bedford Forrest: In Search of the Enigma.* He is the father of two daughters and makes his home in Arizona.

Made in United States
Cleveland, OH
04 November 2026

47649177R00200